Introduction to Seamounts

Contents

Chapter 1

Introduction to Seamounts

1.1 Seamount

For active seamounts, see Submarine volcano.

A **seamount** is a mountain rising from the ocean seafloor that does not reach to the water's surface (sea level), and thus is not an island. Seamounts are typically formed from extinct volcanoes that rise abruptly and are usually found rising from the seafloor to 1,000–4,000 metres (3,300–13,100 ft) in height. They are defined by oceanographers as independent features that rise to at least 1,000 metres (3,281 ft) above the seafloor, characteristically of conical form.*[1] The peaks are often found hundreds to thousands of meters below the surface, and are therefore considered to be within the deep sea.*[2] During their evolution over geologic time, the largest seamounts may reach the sea surface where wave action erodes the summit to form a flat surface. After they have subsided and sunk below the sea surface such flat-top seamounts are called "guyots" or "tablemounts" *[1]

A total of 9,951 seamounts and 283 guyots, covering a total area of 8,796,150 km2 have been mapped*[3] but only a few have been studied in detail by scientists. Seamounts and guyots are most abundant in the North Pacific Ocean, and follow a distinctive evolutionary pattern of eruption, build-up, subsidence and erosion. In recent years, several active seamounts have been observed, for example Loihi in the Hawaiian Islands.

Because of their abundance, seamounts are one of the most common oceanic ecosystems in the world. Interactions between seamounts and underwater currents, as well as their elevated position in the water, attract plankton, corals, fish, and marine mammals alike. Their aggregational effect has been noted by the commercial fishing industry, and many seamounts support extensive fisheries. There are ongoing concerns on the negative impact of fishing on seamount ecosystems, and well-documented cases of stock decline, for example with the orange roughy (*Hoplostethus atlanticus*). 95% of ecological damage is done by bottom trawling, which scrapes whole ecosystems off seamounts.

Because of their large numbers, many seamounts remain to be properly studied, and even mapped. Bathymetry and satellite altimetry are two technologies working to close the gap. There have been instances where naval vessels have collided with uncharted seamounts; for example, Muirfield Seamount is named after the ship that struck it in 1973. However, the greatest danger from seamounts are flank collapses; as they get older, extrusions seeping in the seamounts put pressure on their sides, causing landslides that have the potential to generate massive tsunamis.

1.1.1 Geography

Seamounts can be found in every ocean basin in the world, distributed extremely widely both in space and in age. A seamount is technically defined as an isolated rise in elevation of 1,000 m (3,281 ft) or more from the surrounding seafloor, and with a limited summit area,*[4] of conical form.*[1] If small knolls, ridges and hills less than 1,000 m in height are included there are over 100,000 seamounts in the world ocean.*[3]

Most seamounts are volcanic in origin, and thus tend to be found on oceanic crust near mid-ocean ridges, mantle plumes, and island arcs. Overall, seamount and guyot coverage is greatest as a proportion of seafloor area in the North Pacific Ocean, equal to 4.39% of that ocean region. The Arctic Ocean has only 16 seamounts and no guyots, and the Mediterranean and Black Seas together have only 23 seamounts and 2 guyots. The 9,951 seamounts mapped cover an area of 8,088,550 km2. Seamounts have an average area of 790 km2, with the smallest seamounts found in the Arctic Ocean and the Mediterranean and Black Seas, whilst the largest mean seamount size occurs in the Indian Ocean (890 km2). The largest seamount has an area of 15,500 km2 and it occurs in the North Pacific. Guyots cover a total area of 707,600 km2 and have an average area of 2,500 km2, more than twice the average size of seamounts. Nearly 50% of guyot area and 42% of the number of guyots occur in the North Pacific Ocean, covering 342,070 km2. The largest three guyots are all in the North Pacific: the Kuko Guyot (estimated 24,600 km2), Suiko Guyot (estimated 20,220 km2) and the Pallada Guyot (estimated 13,680 km2).[3]

Grouping

"Seamount chain" redirects here; for a broader coverage related to this topic, see Undersea mountain range.

Seamounts are often found in groupings or submerged archipelagos, a classic example being the Emperor Seamounts, an extension of the Hawaiian Islands. Formed millions of years ago by volcanism, they have since subsided far below sea level. This long chain of islands and seamounts extends thousands of kilometers northwest from the island of Hawaii.

Distribution of seamounts and guyots in the North Pacific

There are more seamounts in the Pacific Ocean than in the Atlantic, and their distribution can be described as comprising several elongate chains of seamounts superimposed on a more or less random background distribution.[5] Seamount chains occur in all three major ocean basins, with the Pacific having the most number and most extensive seamount chains. These include the Hawaiian (Emperor), Mariana, Gilbert, Tuomotu and Austral Seamounts (and island groups) in the north Pacific and the Louisville and Sala y Gomez ridges in the southern Pacific Ocean. In the North Atlantic Ocean, the New England Seamount Chain extends from the eastern coast of the United States to the mid-ocean ridge. Craig and Sandwell[5] noted that clusters of larger Atlantic seamounts tend to be associated with other evidence of hotspot activity,

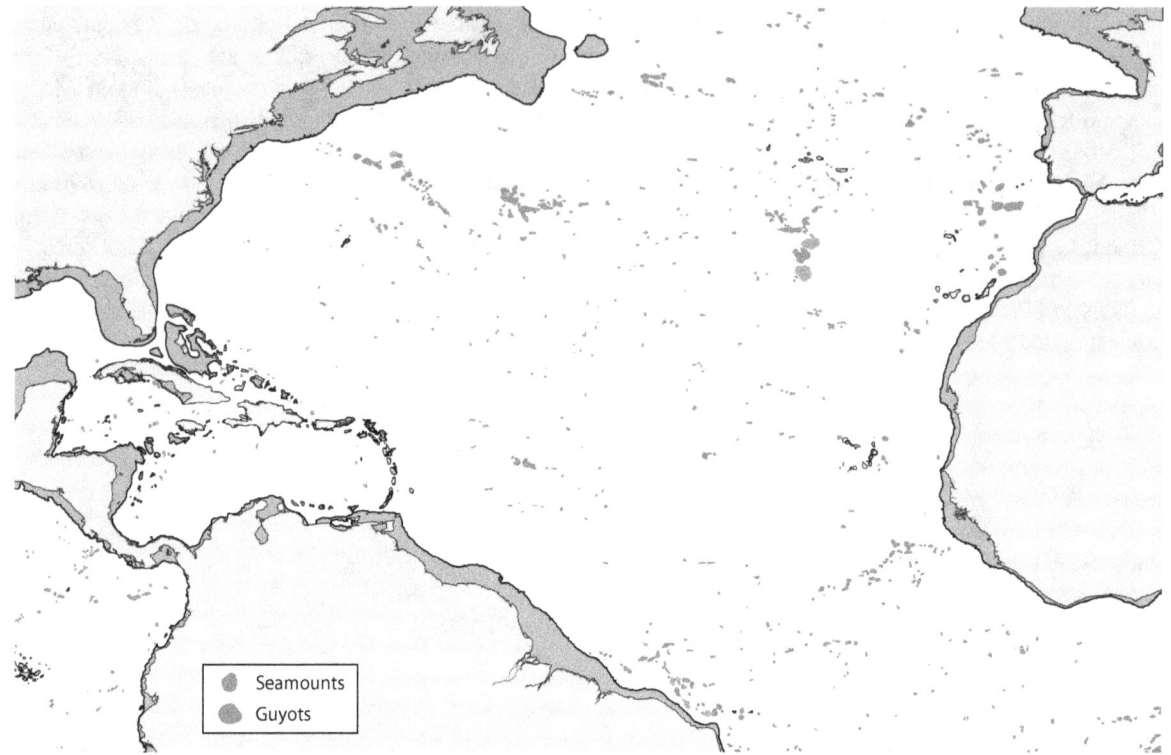

Distribution of seamounts and guyots in the North Atlantic

such as on the Walvis Ridge, Bermuda Islands and Cape Verde Islands. The mid-Atlantic ridge and spreading ridges in the Indian Ocean are also associated with abundant seamounts.[6] Otherwise, seamounts tend not to form distinctive chains in the Indian and Southern Oceans, but rather their distribution appears to be more or less random.

Isolated seamounts and those without clear volcanic origins are less common; examples include Bollons Seamount, Eratosthenes Seamount, Axial Seamount and Gorringe Ridge.[7] If all known seamounts were collected into one area, they would make a landform the size of Europe.[8] Their overall abundance makes them one of the most common, and least understood, marine structures and biomes on Earth,[9] a sort of exploratory frontier.[10]

1.1.2   Geology

Geochemistry and evolution

Most seamounts are built by one of two volcanic processes, although some, such as the Christmas Island Seamount Province near Australia, are more enigmatic.[11] Volcanoes near plate boundaries and mid-ocean ridges are built by decompression melting of rock in the mantle that then floats up to the surface, while volcanoes formed near subducting zones are created because the subducting plate adds volatiles to the rising plate that lowers its melting point. Which process formed the seamount has a profound effect on its eruptive materials. Lava flows from mid-ocean ridge and plate boundary seamounts are mostly basaltic (both tholeiitic and alkalic), whereas flows from subducting ridge volcanoes are mostly calc-alkaline lavas. Compared to mid-ocean ridge seamounts, subduction zone seamounts generally have more sodium, alkali, and volatile abundances, and less magnesium, resulting in more explosive, viscous eruptions.[10]

All volcanic seamounts follow a particular pattern of growth, activity, subsidence and eventual extinction. The first stage of a seamount's evolution is its early activity, building its flanks and core up from the sea floor. This is followed by a period of intense volcanism, during which the new volcano erupts almost all (e.g. 98%) of its total magmatic volume. The seamount may even grow above sea level to become an oceanic island (for example, the 2009 eruption of Hunga Tonga). After a period of explosive activity near the ocean surface, the eruptions slowly die away. With eruptions becoming

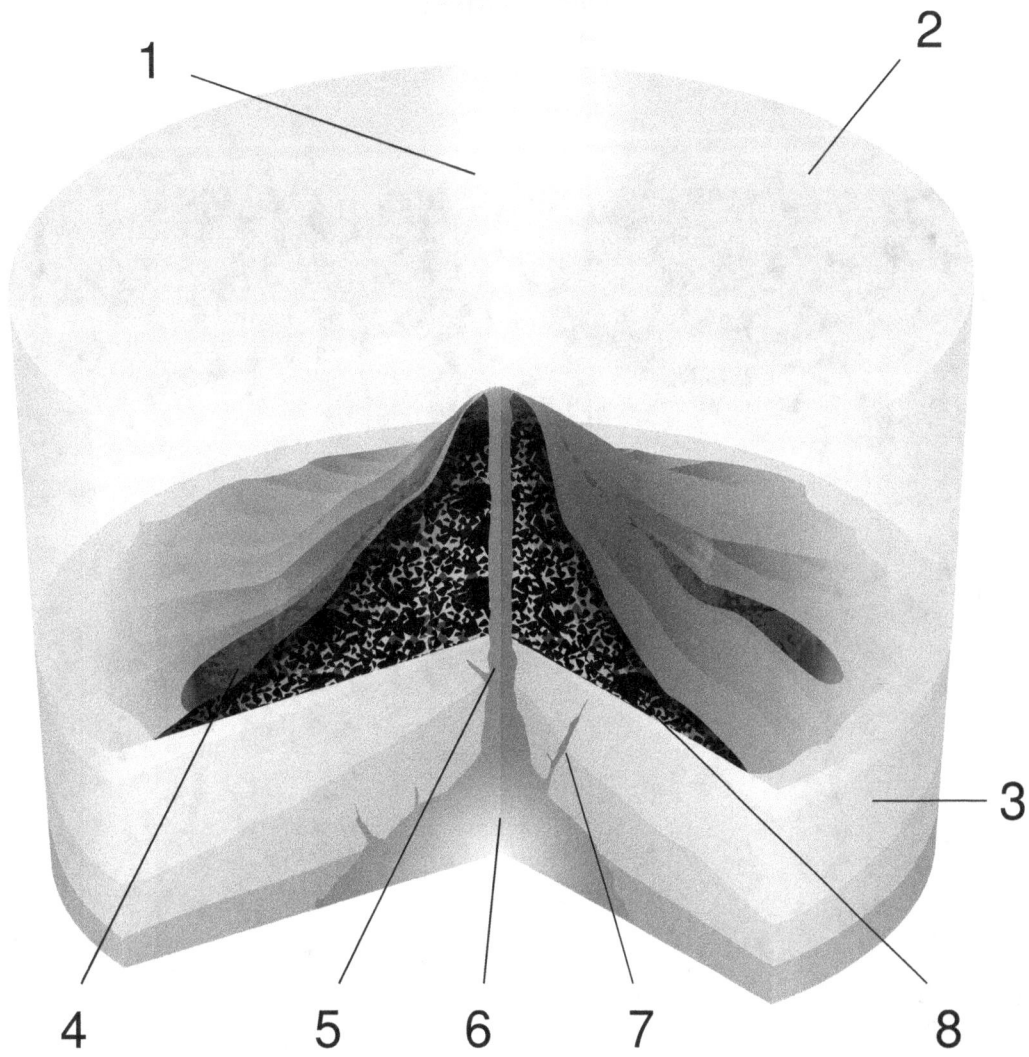

Diagram of a submarine eruption. (key: 1. Water vapor cloud 2. Water 3. Stratum 4. Lava flow 5. Magma conduit 6. Magma chamber 7. Dike 8. Pillow lava) Click to enlarge.

infrequent and the seamount losing its ability to maintain itself, the volcano starts to erode. After finally becoming extinct (possibly after a brief rejuvenated period), they are ground back down by the waves. Seamounts are built in a far more dynamic oceanic setting than their land counterparts, resulting in horizontal subsidation as the seamount moves on the grinding plate towards a subduction zone. Here it is subducted under the plate margin and ultimately destroyed, but it may leave evidence of its passage by carving an indentation into the opposing wall of the subduction trench. The majority of seamounts have already completed their eruptive cycle, so access to early flows by researchers is limited by late volcanic activity.*[10]

Ocean-ridge volcanoes in particular have been observed to follow a certain pattern in terms of eruptive activity, first observed with Hawaiian seamounts but now shown to be the process followed by all seamounts of the ocean-ridge type. During the first stage the volcano erupts basalt of various types, caused by various degrees of mantle melting. In the second, most active stage of its life, ocean-ridge volcanoes erupt tholeiitic to mildly alkalic basalt as a result of a larger area melting in the mantle. This is finally capped by alkalic flows late in its eruptive history, as the link between the seamount and its source of volcanism is cut by crustal movement. Some seamounts also experience a brief "rejuvenated" period after a hiatus of 1.5 to 10 million years, the flows of which are highly alkalic and produce many xenoliths.[10]

In recent years, geologists have confirmed that a number of seamounts are active undersea volcanoes; two examples are Lo 'ihi in the Hawaiian Islands and Vailulu'u in the Manu'a Group (Samoa).[7]

Lava types

Pillow lava, a type of basalt flow that originates from lava-water interactions during submarine eruptions.[12]

The most apparent lava flows at a seamount are the eruptive flows which cover their flanks, however igneous intrusions, in the forms of dikes and sills, are also an important part of seamount growth. The most common type of flow is pillow lava, named so after its unusual shape. Less common are sheet flows, which are glassy and marginal, and indicative of larger-scale flows. Volcaniclastic sedimentary rocks dominate shallow-water seamounts. They are the products of the explosive activity of seamounts that are near the water's surface, and can also form from mechanical wear of existing volcanic rock.[10]

Structure

Seamounts can form in a wide variety of tectonic settings, resulting in a very diverse structural bank. Seamounts come in a wide variety of structural shapes, from conical to flat-topped to complexly shaped.[10] Some are built very large and very low, such as Koko Guyot[13] and Detroit Seamount;[14] others are built more steeply, such as Loihi Seamount[15] and Bowie Seamount.[16] Some seamounts also have a carbonate or sediment cap.[10]

Many seamounts show signs of intrusive activity, which is likely to lead to inflation, steepening of volcanic slopes, and ultimately, flank collapse.[10] There are also several sub-classes of seamounts. The first are guyots, seamounts with a flat top. These tops must be 200 m (656 ft) or more below the surface of the sea; the diameters of these flat summits can be over 10 km (6.2 mi).[17] Knolls are isolated elevation spikes measuring less than 1,000 meters (3,281 ft). Lastly, pinnacles are small pillar-like seamounts.[4]

1.1.3 Ecology

Ecological role of seamounts

Seamounts are exceptionally important to their biome ecologically, but their role in their environment is poorly understood. Because they project out above the surrounding sea floor, they disturb standard water flow, causing eddies and associated hydrological phenomena that ultimately result in water movement in an otherwise still ocean bottom. Currents have been measured at up to 0.9 knots, or 48 centimeters per second. Because of this upwelling seamounts often carry above-average plankton populations, seamounts are thus centers where the fish that feed on them aggregate, in turn falling prey to further predation, making seamounts important biological hotspots.[4]

Seamounts provide habitats and spawning grounds for these larger animals, including numerous fish. Some species, including black oreo *(Allocyttus niger)* and blackstripe cardinalfish *(Apogon nigrofasciatus)*, have been shown to occur more often on seamounts than anywhere else on the ocean floor. Marine mammals, sharks, tuna, and cephalopods all congregate over seamounts to feed, as well as some species of seabirds when the features are particularly shallow.[4]

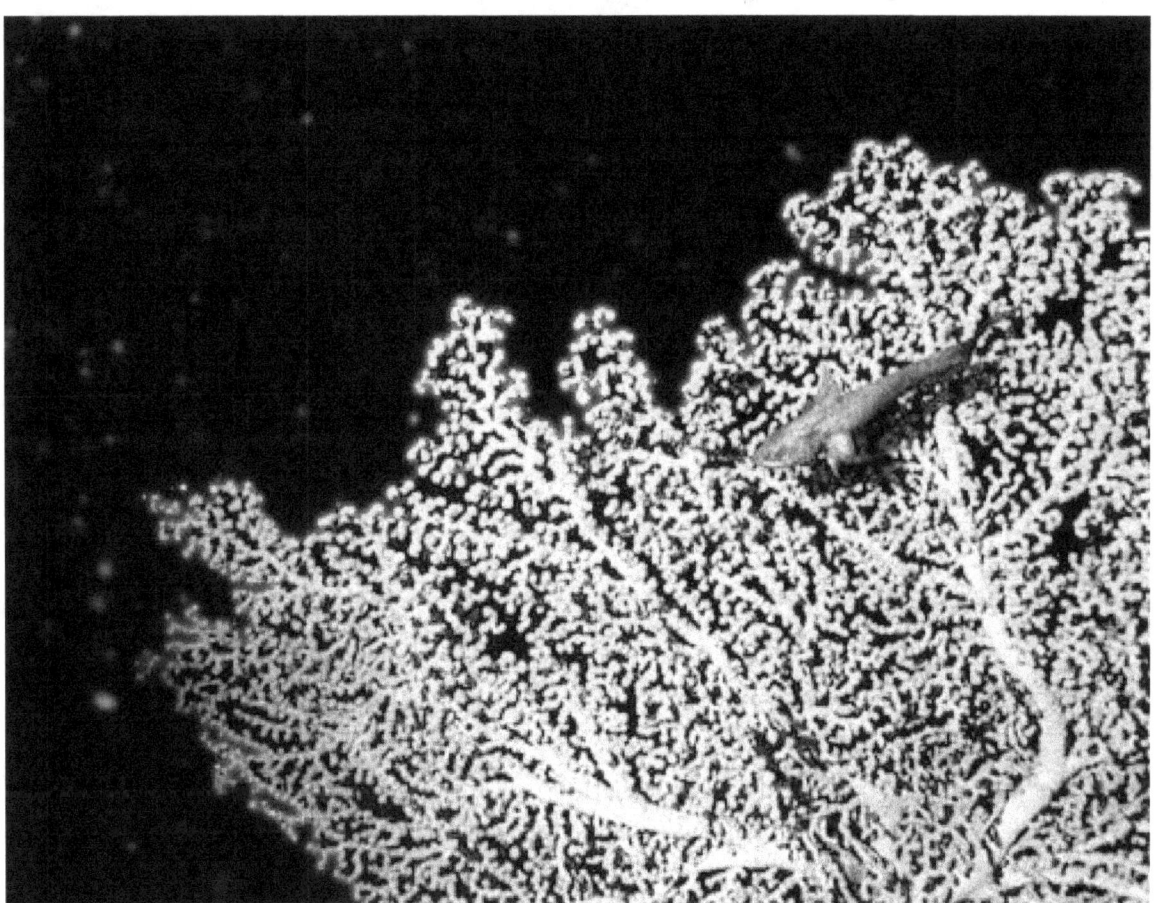

Grenadier fish (Coryphaenoides sp.*) and bubblegum coral (*Paragorgia arborea*) on the crest of Davidson Seamount. These are two species attracted to the seamount;* Paragorgia arborea *in particular grows in the surrounding area as well, but nowhere near as profusely.* [18]

Seamounts often project upwards into shallower zones more hospitable to sea life, providing habitats for marine species that are not found on or around the surrounding deeper ocean bottom. Because seamounts are isolated from each other they form "undersea islands" creating the same biogeographical interest. As they are formed from volcanic rock, the substrate is much harder than the surrounding sedimentary deep sea floor. This causes a different type of fauna to exist than on the seafloor, and leads to a theoretically higher degree of endemism.[19] However, recent research especially centered at Davidson Seamount suggests that seamounts may not be especially endemic, and discussions are ongoing on the effect of seamounts on endemicity. They *have*, however, been confidently shown to provide a habitat to species that have difficulty surviving elsewhere.[20][21]

The volcanic rocks on the slopes of seamounts are heavily populated by suspension feeders, particularly corals, which capitalize on the strong currents around the seamount to supply them with food. This is in sharp contrast with the typical deep-sea habitat, where deposit-feeding animals rely on food they get off the ground.[4] In tropical zones extensive coral growth results in the formation of coral atolls late in the seamount's life.[21][22]

In addition soft sediments tend to accumulate on seamounts, which are typically populated by polychaetes (annelid marine worms) oligochaetes (microdrile worms), and gastropod mollusks (sea slugs). Xenophyophores have also been found. They tend to gather small particulates and thus form beds, which alters sediment deposition and creates a habitat for smaller animals.[4] Many seamounts also have hydrothermal vent communities, for example Suiyo[23] and Loihi seamounts.[24] This is helped by geochemical exchange between the seamounts and the ocean water.[10]

Seamounts may thus be vital stopping points for some migratory animals, specifically whales. Some recent research indicates whales may use such features as navigational aids throughout their migration.[25] For a long time it has been surmised that many pelagic animals visit seamounts as well, to gather food, but proof of this aggregating effect has been lacking. The first demonstration of this conjecture was published in 2008.[26]

Fishing

The effect that seamounts have on fish populations has not gone unnoticed by the commercial fishing industry. Seamounts were first extensively fished in the second half of the 20th century, due to poor management practices and increased fishing pressure seriously depleting stock numbers on the typical fishing ground, the continental shelf. Seamounts have been the site of targeted fishing since that time.[27]

Nearly 80 species of fish and shellfish are commercially harvested from seamounts, including spiny lobster (**Palinuridae**), mackerel (**Scombridae** and others), red king crab (*Paralithodes camtschaticus*), red snapper (*Lutjanus campechanus*), tuna (**Scombridae**), Orange roughy (*Hoplostethus atlanticus*), and perch (**Percidae**).[4]

Conservation

The ecological conservation of seamounts is hurt by the simple lack of information available. Seamounts are very poorly studied, with only 350 of the estimated 100,000 seamounts in the world having received sampling, and fewer than 100 in depth.[28] Much of this lack of information can be attributed to a lack of technology, and to the daunting task of reaching these underwater structures; the technology to fully explore them has only been around the last few decades. Before consistent conservation efforts can begin, the seamounts of the world must first be mapped, a task that is still in progress.[4]

Overfishing is a serious threat to seamount ecological welfare. There are several well-documented cases of fishery exploitation, for example the orange roughy (*Hoplostethus atlanticus*) off the coasts of Australia and New Zealand and the pelagic armorhead (*Pseudopentaceros richardsoni*) near Japan and Russia.[4] The reason for this is that the fishes that are targeted over seamounts are typically long-lived, slow-growing, and slow-maturing. The problem is confounded by the dangers of trawling, which damages seamount surface communities, and the fact that many seamounts are located in international waters, making proper monitoring difficult.[27] Bottom trawling in particular is extremely devastating to seamount ecology, and is responsible for as much as 95% of ecological damage to seamounts.[29]

Corals from seamounts are also vulnerable, as they are highly valued for making jewellery and decorative objects. Significant harvests have been produced from seamounts, often leaving coral beds depleted.[4]

Individual nations are beginning to note the effect of fishing on seamounts, and the European Commission has agreed to fund the OASIS project, a detailed study of the effects of fishing on seamount communities in the North Atlantic.[27]

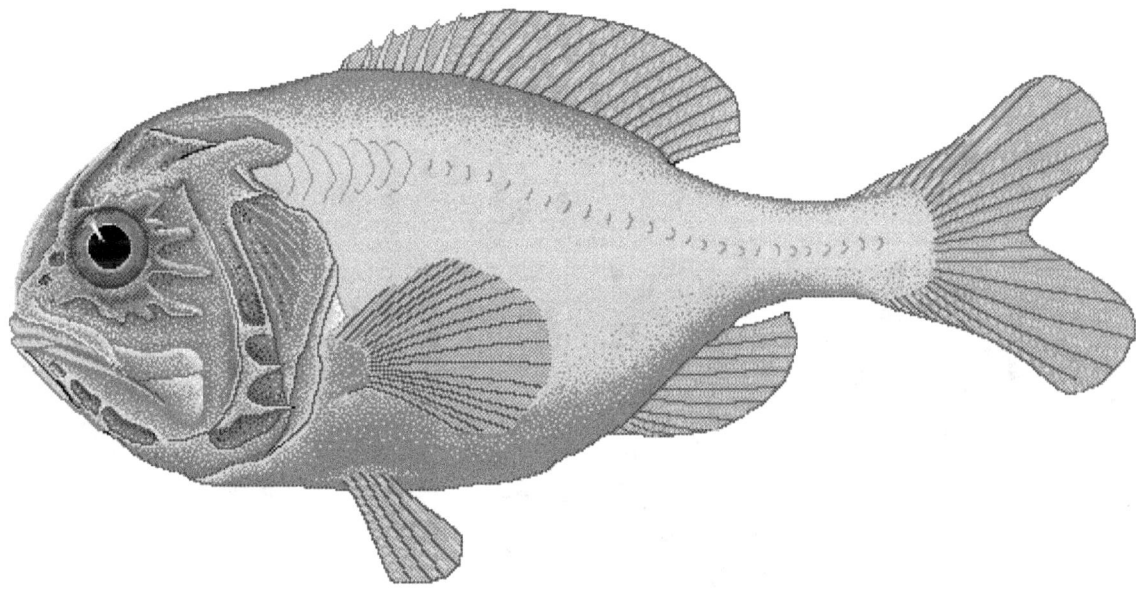

Because of overfishing at their seamount spawning grounds, stocks of orange roughy (Hoplostethus atlanticus) *have plummeted; experts say that it could take decades for the species to restore itself to its former numbers.*[27]

Coral earrings of this type are often made from coral harvested off seamounts.

Another project working towards conservation is CenSeam, a Census of Marine Life project formed in 2005. CenSeam is intended to provide the framework needed to prioritise, integrate, expand and facilitate seamount research efforts in order to significantly reduce the unknown and build towards a global understanding of seamount ecosystems, and the roles they have in the biogeography, biodiversity, productivity and evolution of marine organisms.*[28]*[30]

Possibly the best ecologically studied seamount in the world is Davidson Seamount, with six major expeditions recording over 60,000 species observations. The contrast between the seamount and the surrounding area was well-marked.*[20] One of the primary ecological havens on the seamount is its deep sea coral garden, and many of the specimens noted were over a century old.*[18] Following the expansion of knowledge on the seamount there was extensive support to make it a marine sanctuary, a motion that was granted in 2008 as part of the Monterey Bay National Marine Sanctuary.*[31] Much of what is known about seamounts ecologically is based on observations from Davidson.*[18]*[26] Another such seamount is Bowie Seamount, which has also been declared a marine protected area by Canada for its ecological richness.[*][32]

1.1.4 Exploration

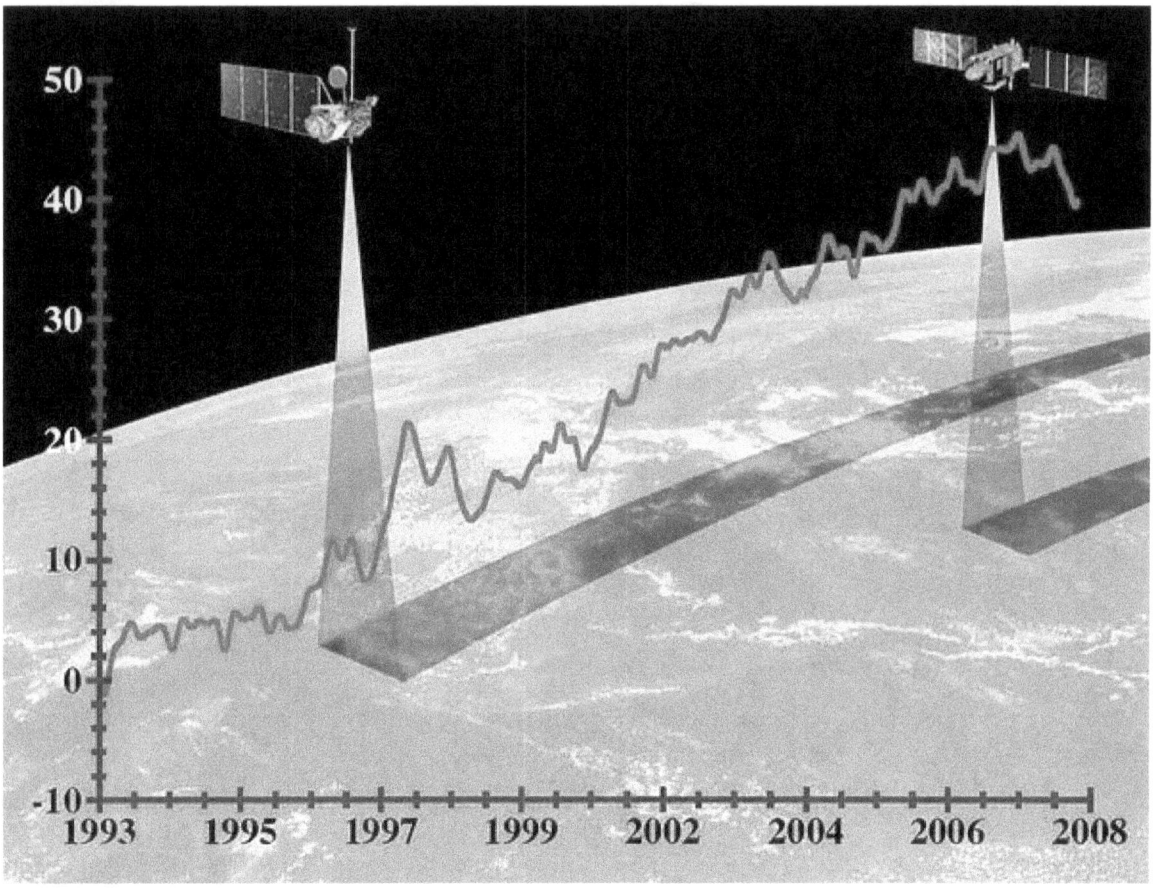

Graph showing the rise in global sea level (in mm) as measured by the NASA/CNES oceanic satellite altimeter TOPEX/Poseidon (left) and its follow-on mission Jason-1.

The study of seamounts has been stymied for a long time by the lack of technology. Although seamounts have been sampled as far back as the 19th century, their depth and position meant that the technology to explore and sample seamounts in sufficient detail did not exist until the last few decades. Even with the right technology available, only a scant 1% of the total number have been explored,[*][8] and sampling and information remains biased towards the top 500 m (1,640 ft).[*][4] New species are observed or collected and valuable information is obtained on almost every submersible dive at seamounts.[*][9]

Before seamounts and their oceanographic impact can be fully understood, they must be mapped, a daunting task due to their sheer number.[*][4] The most detailed seamount mappings are provided by multibeam echosounding (sonar), however after more than 5000 publicly held cruises, the amount of the sea floor that has been mapped remains minuscule. Satellite altimetry is a broader alternative, albeit not as detailed, with 13,000 catalogued seamounts; however this is still only a fraction of the total 100,000. The reason for this is that uncertainties in the technology limit recognition to features 1,500 m (4,921 ft) or larger. In the future, technological advances could allow for a larger and more detailed catalogue.[*][22]

Data from CryoSat-2 has shown 25,000 seamounts, with more to come as data is interpreted.[*][33][*][34][*][35][*][36]

1.1.5 Deep-sea mining

Seamounts are a possible future source of heavy metals. The growth of the human population and, with it, heavy industry, has pressed demands for Earth's finite resources. Even though the ocean makes up 70% of the world, technological challenges with deep-sea mineral mining have severely limited its extent. But with the constantly decreasing supply on land, many see oceanic mining as the destined future, and seamounts stand out as candidates.[*][37]

Seamounts are abundant, and all have metal resource potential because of various enrichment processes during the seamount's life. An example for epithermal gold mineralization on the seafloor is Conical Seamount, located about 8km south of Lihir Island in Papua New Guinea. Conical Seamount has a basal diameter of about 2.8km and rises about 600m above the seafloor to a water depth of 1050m. Grab samples from its summit contain the highest gold concentrations yet reported from the modern seafloor (max. 230g/t Au, avg. 26g/t, n=40).[*][38] Hydrogenic Iron-manganese, hydrothermal iron oxide, sulfide, sulfate, sulfur, hydrothermal manganese oxide, and phosphorite[*][39] (the latter especially in parts of Micronesia) are all mineral resources that are founded by various processes and deposited upon seamounts. However, only the first two have any potential of being targeted by mining in the next few decades.[*][37]

1.1.6 Dangers

USS San Francisco *in dry dock in Guam in January 2005, following its collision with an uncharted seamount. The damage was extensive and the submarine was just barely salvaged.[*][40]*

See also: Landslide § Causing tsunamis and Landslide § Prehistoric submarine landslides

Some seamounts have not been mapped and thus pose a navigational danger. For instance, Muirfield Seamount is named after the ship that hit it in 1973.[*][41] More recently, the submarine USS *San Francisco* ran into an uncharted seamount in 2005 at a speed of 35 knots (40.3 mph; 64.8 km/h), sustaining serious damage and killing one seaman.[*][40]

One major seamount risk is that often, in the late of stages of their life, extrusions begin to seep in the seamount. This activity leads to inflation, over-extension of the volcano's flanks, and ultimately flank collapse, leading to submarine landslides with the potential to start major tsunamis, which can be among the largest natural disasters in the world. In an illustration of the potent power of flank collapses, a summit collapse on the northern edge of Vlinder Seamount resulted in a pronounced headwall scarp and a field of debris up to 6 km (4 mi) away.[*][10] A catastrophic collapse at Detroit Seamount flattened its whole structure extensively.[*][14] Lastly, in 2004, scientists found marine fossils 61 m (200 ft) up the flank of Kohala mountain in Hawaii (island). Subsidence analysis found that at the time of their deposition, this would have been 500 m (1,640 ft) up the flank of the volcano,[*][42] far too high for a normal wave to reach. The date corresponded with a massive flank collapse at the nearby Mauna Loa, and it was theorized that it was a massive tsunami, generated by the landslide, that deposited the fossils.[*][43]

1.1.7 See also

1.1.8 References

[1] IHO, 2008. Standardization of Undersea Feature Names: Guidelines Proposal form Terminology, 4th ed. International Hydrographic Organisation and Intergovernmental Oceanographic Commission, Monaco.

[2] Nybakken, James W. and Bertness, Mark D., 2008. *Marine Biology: An Ecological Approach*. Sixth Edition. Benjamin Cummings, San Francisco

[3] Harris, P.T., MacMillan-Lawler, M., Rupp, J., Baker, E.K., 2014. Geomorphology of the oceans. Marine Geology 352, 4-24

[4] "Seamount". *Encyclopedia of Earth*. December 9, 2008. Retrieved 24 July 2010.

[5] Craig, C.H., Sandwell, D.T., 1988. Global distribution of seamounts from Seasat profiles. Journal of Geophysical Research 93, 10,408-410,420.

[6] Kitchingman, A., Lai, S., 2004. Inferences on Potential Seamount Locations from Mid-Resolution Bathymetric Data. in: Morato, T., Pauly, D. (Eds.), FCRR Seamounts: Biodiversity and Fisheries. Fisheries Centre Research Reports. University of British Columbia, Vanvouver, BC, pp. 7-12.

[7] Keating, B.H.; et al. (1987). *Geophysical Monogram: Seamounts, islands and atolls*. American Geophysical Union. ISBN 978-0-87590-068-1. ISSN 0065-8448. Retrieved 24 July 2010.

[8] "Seamount Scientists Offer New Comprehensive View of Deep-Sea Mountains". ScienceDaily. 23 February 2010. Retrieved 25 July 2010.

[9] "Seamounts Identified as Significant, Unexplored Territory". ScienceDirect. 30 April 2010. Retrieved 25 July 2010.

[10] Hubert Straudigal and David A Clague. "The Geological History of Deep-Sea Volcanoes: Biosphere, Hydrosphere, and Lithosphere Interactions" (PDF). *Oceanography*. Seamounts Special Issue (Oceanography Society) **32** (1). Retrieved 25 July 2010.

[11] K. Hoernle, F. Hauff, R. Werner, P. van den Bogaard, A. D. Gibbons, S. Conrad, and R. D. Müller (27 November 2011). "Origin of Indian Ocean Seamount Province by shallow recycling of continental lithosphere". *Nature Geoscience* (Nature Publishing Group) **4**: 883–887. Bibcode:2011NatGe...4..883H. doi:10.1038/ngeo1331. Retrieved 30 December 2011.

[12] "Pillow lava". NOAA. Retrieved 25 July 2010.

[13] "SITE 1206". *Ocean Drilling Program Database-Results of Site 1206*. Ocean Drilling Program. Retrieved 26 July 2010.

[14] Kerr, B. C., D. W. Scholl, and S. L. Klemperer (July 12, 2005). "Seismic stratigraphy of Detroit Seamount, Hawaiian–Emperor Seamount chain" (PDF). Stanford University. Retrieved 15 July 2010.

[15] Rubin, Ken (January 19, 2006). "General Information About Loihi". *Hawaii Center for Volcanology*. SOEST. Retrieved 26 July 2010.

[16]   "The Bowie Seamount Area" (PDF). John F. Dower and Frances J. Fee. February 1999. Retrieved 26 July 2010.

[17]   "Guyots". Encyclopædia Britannica Online. Retrieved 24 July 2010.

[18]   "Seamounts may serve as refuges for deep-sea animals that struggle to survive elsewhere". PhysOrg. February 11, 2009. Retrieved December 7, 2009.

[19]   "Davidson Seamount" (PDF). NOAA, Monterey Bay National Marine Sanctuary. 2006. Retrieved 2 December 2009.

[20]   McClain, Craig R.; Lundsten L., Ream M., Barry J., DeVogelaere A. (January 7, 2009). Rands, Sean, ed. "Endemicity, Biogeography, Composition, and Community Structure On a Northeast Pacific Seamount". *PLoS ONE* **1** (4): e4141. Bibcode:2009PLoSO...4.4141M. doi:10.1371/journal.pone.0004141. PMC 2613552. PMID 19127302. Retrieved December 3, 2009.

[21]   Lundsten, L; J. P. Barry, G. M. Cailliet, D. A. Clague, A. DeVogelaere, J. B. Geller (January 13, 2009). "Benthic invertebrate communities on three seamounts off southern and central California". *Marine Ecology Progress Series* (Inter-Research Science Center) **374**: 23–32. doi:10.3354/meps07745.

[22]   Pual Wessel, David T. Sandwell, Seung-Sep Kim. "The Global Seamount Census" (PDF). *Oceanography*. Seamounts Special Issue (Oceanography Society) **23** (1). ISSN 1042-8275. Retrieved 25 June 2010.

[23]   Higashi, Y; et al. (15 March 2004). "Microbial diversity in hydrothermal surface to subsurface environments of Suiyo Seamount, Izu-Bonin Arc, using a catheter-type in situ growth chamber". *FEMS Microbiology Ecology* **47** (3): 327–336. doi:10.1016/S0168-6496(04)00004-2. PMID 19712321. Retrieved 25 July 2010.

[24]   "Introduction to the Biology and Geology of Lōʻihi Seamount". *Lōʻihi Seamount*. Fe-Oxidizing Microbial Observatory (FeMO). 2009-02-01. Retrieved 2009-03-02.

[25]   Kennedy, Jennifer. "Seamount: What is a Seamount?". ask.com. Retrieved 25 July 2010.

[26]   Morato, T., Varkey, D.A., Damaso, C., Machete, M., Santos, M., Prieto, R., Santos, R.S. and Pitcher, T.J. (2008). "Evidence of a seamount effect on aggregating visitors". *Marine Ecology Progress* Series 357: 23-32.

[27]   "Seamounts – hotspots of marine life". International Council for the Exploration of the Sea. Retrieved 24 July 2010.

[28]   "CenSeam Mission". CenSeam. Archived from the original on 24 May 2010. Retrieved 22 July 2010.

[29]   Report of the Secretary-General (2006) *The Impacts of Fishing on Vulnerable Marine Ecosystems* United Nations. 14 July 2006. Retrieved on 26 July 2010.

[30]   "CenSeam Science". CenSeam. Retrieved 22 July 2010.

[31]   "NOAA Releases Plans for Managing and Protecting Cordell Bank, Gulf of Farallones and Monterey Bay National Marine Sanctuaries" (PDF). *Press release*. NOAA. November 20, 2008. Retrieved 2 December 2009.

[32]   "Bowie Seamount Marine Protected Area". Fisheries and Oceans Canada. 1 October 2011. Retrieved 31 December 2011.

[33]   Amos, Jonathan. "Satellites detect 'thousands' of new ocean-bottom mountains" *BBC News*, 2 October 2014.

[34]   "New Map Exposes Previously Unseen Details of Seafloor"

[35]   David T. Sandwell, R. Dietmar Müller, Walter H. F. Smith, Emmanuel Garcia, Richard Francis. "New global marine gravity model from CryoSat-2 and Jason-1 reveals buried tectonic structure" *Science* 3 October 2014: Vol. 346 no. 6205 pp. 65-67. DOI: 10.1126/science.1258213

[36]   "Cryosat 4 Plus" *DTU Space*

[37]   James R. Hein, Tracy A. Conrad, Hubert Staudigel. "Seamount Mineral Deposits: A Source for Rare Minerals for High Technology Industries" (PDF). *Oceanography*. Seamounts Special Issue (Oceanography Society) **23** (1). ISSN 1042-8275. Retrieved 26 July 2010.

[38]   Muller, Daniel; Leander Franz; Sven Petersen; Peter Herzig; Mark Hannington (2003). "Comparison between magmatic activity and gold mineralization at Conical Seamount and Lihir Island, Papua New Guinea". *Mineralogy and Petrology* **79**: 259–283.

[39] C.Michael Hogan. 2011. *Phosphate*. Encyclopedia of Earth. Topic ed. Andy Jorgensen. Ed.-in-Chief C.J.Cleveland. National Council for Science and the Environment. Washington DC

[40] "USS San Francisco (SSN 711)". Retrieved 25 July 2010.

[41] Nigel Calder (2002). *How to Read a Navigational Chart: A Complete Guide to the Symbols, Abbreviations, and Data Displayed on Nautical Charts*. International Marine/Ragged Mountain Press.

[42] Seach, John. "Kohala Volcano". *Volcanism reference base*. John Seach, vulcanologist. Retrieved 25 July 2010.

[43] "Hawaiian tsunami left a gift at foot of volcano". *New Scientist* (Reed Business Information) (2464): 14. 2004-09-11. Retrieved 25 July 2010.

1.1.9 Bibliography

Geology

- Keating, B.H., Fryer, P., Batiza, R., Boehlert, G.W. (Eds.), 1987: *Seamounts, islands and atolls*. Geophys. Monogr. 43:319-334.

- Menard, H.W. (1964). *Marine Geology of the Pacific*. International Series in the Earth Sciences. McGraw-Hill, New York, 271 pp.

Ecology

- Clark, M. R.; Rowden, A. A.; Schlacher, T.; Williams, A.; Consalvey, M.; Stocks, K. I.; Rogers, A. D.; O'Hara, T. D.; White, M.; Shank, T. M.; Hall-Spencer, J. M. (2010). "The Ecology of Seamounts: Structure, Function, and Human Impacts". *Annual Review of Marine Science* **2**: 253–278. doi:10.1146/annurev-marine-120308-081109. PMID 21141665.

- Richer de Forges, J. Anthony Koslow, and G. C. B. Poore (22 June 2000). "Diversity and endemism of the benthic seamount fauna in the southwest Pacific". *Nature* (Nature Publishing Group) **405** (6789): 944–947. doi:10.1038/35016066. PMID 10879534.

- Koslow, J.A. (1997). *Seamounts and the ecology of deep-sea fisheries*. Am. Sci. 85:168-176.

- Lundsten L, McClain CR, Barry JP, Cailliet GM, Clague DA, DeVogelaere AP (2009) Ichthyofauna on Three Seamounts off Southern and Central California, USA. Marine Ecology Progress Series 389:223-232.

- Pitcher, T.J., Morato, T., Hart, P.J.B., Clark, M.R., Haggan, N. and Santos, R.S. (eds) (2007). "Seamounts: Ecology, Fisheries and Conservation". *Fish and Aquatic Resources* Series 12, Blackwell, Oxford, UK. 527pp. ISBN 978-1-4051-3343-2

1.1.10 External links

Geography and geology

- Earthref Seamount Catalogue. A database of seamount maps and catalogue listings.

- Volcanic History of Seamounts in the Gulf of Alaska.

- The giant Ruatoria debris avalanche on the northern Hikurangi margin, New Zealand. Aftermath of a seamount carving into the far side of a subduction trench.

- Evolution of Hawaiian volcanoes. The life cycle of seamounts was originally observed off of the Hawaiian arc.

- How Volcanoes Work: Lava and Water. An explanation of the different types of lava-water interactions.

Ecology

- A review of the effects of seamounts on biological processes. NOAA paper.

- Mountains in the Sea, a volume on the biological and geological effects of seamounts, available fully online.

- SeamountsOnline, seamount biology database.

- Vulnerability of deep sea corals to fishing on seamounts beyond areas of national jurisdiction, United Nations Environment Program.

1.2   Guyot

For other uses, see Guyot (disambiguation).

A **guyot** /giːˈjoʊ/, also known as a **tablemount**, is an isolated underwater volcanic mountain (seamount), with a flat

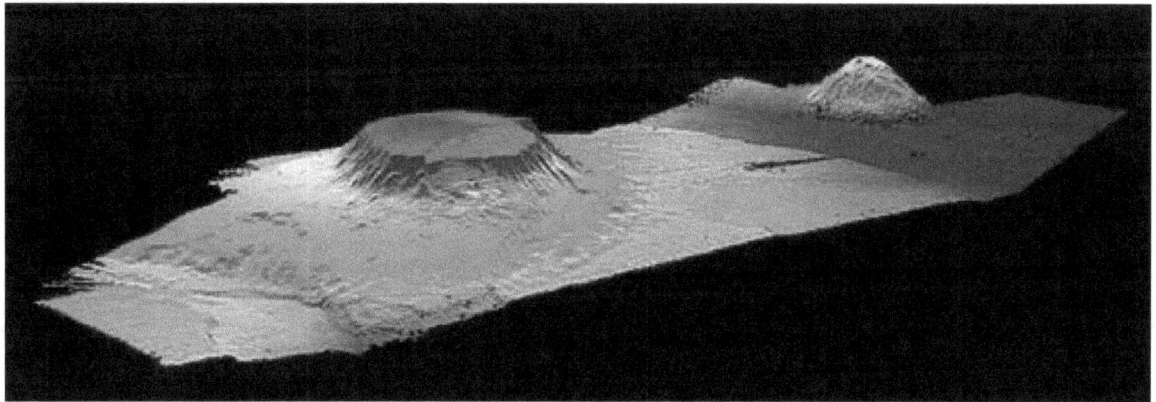

The Bear Seamount, a Guyot

top over 200 metres (660 feet) below the surface of the sea. The diameters of these flat summits can exceed 10 km (6.2 mi).*[1]

The guyot was named after the Swiss-American geographer and geologist Arnold Henry Guyot (died 1884). The term was coined by Harry Hammond Hess. Guyots are most commonly found in the Pacific Ocean. Guyots show evidence of having been above the surface with gradual subsidence through stages from fringed reefed mountain, coral atoll, and finally a flat topped submerged mountain.*[1] The steepness gradient of most guyots is about 20 degrees. To technically be considered a guyot or tablemount, they must stand at least 3000 ft (900 m) tall. However, there are many undersea mounts that can range from just less than 300 ft to around 3000 ft. Very large oceanic volcanic constructions, hundreds of km across, are called oceanic plateaus.*[2] Seamounts are made by extrusion of lavas piped upward in stages from sources within the Earth's mantle to vents on the seafloor. Seamounts provide data on movements of tectonic plates on which they ride, and on the rheology of the underlying lithosphere. The trend of a seamount chain traces the direction of motion of the lithospheric plate over a more or less fixed heat source in the underlying asthenosphere part of the Earth's mantle.*[3] There are thought to be an estimated 2,000 seamounts in the Pacific basin. The Emperor Seamounts are an excellent example of an entire volcanic chain undergoing this process and contain many guyots among their other examples.

Another factor contributing to the guyots being underwater has to do with the oceanic ridges, such as the Mid-Atlantic Ridge in the Atlantic Ocean. Mid-ocean ridges gradually spread apart over time, due to molten lava being pushed up under the surface of the earth and creating new rock. As the mid-ocean ridges spread apart, the guyots move with them, thus continually sinking deeper into the depths of the ocean. Thus, the greater amount of time that passes, the deeper the guyots become.*[4] Although guyots can be hundreds of millions of years old, there have been some recently discovered guyots that were only formed within the last 1 million years, including Bowie Seamount on the coast of British Columbia, Canada.

One guyot in particular, the Great Meteor Tablemount in the Northeast Atlantic Ocean, stands at more than 4000 m (13,120 ft). The guyot's diameter is 110 km (68 mi).[*][5] Guyots are also associated with specific lifeforms and varying amounts of organic matter. Local increases in chlorophyll a, enhanced carbon incorporation rates and changes in phytoplankton species composition were associated with the seamount.[*][6]

Guyots were first recognized by Harry Hammond Hess in 1965 who collected data using echo-sounding equipment on a ship he commanded during World War II.[*][7] The data showed the configuration of the seafloor where he saw that some undersea mountains had flat tops. Hess called these undersea mountains 'guyots' because they resembled Guyot Hall, the flat roofed biology and geology building at Princeton University which was itself named after the 18th century geographer Arnold Henry Guyot.[*][8] Hess postulated they were once volcanic islands that were beheaded by wave action yet they are now deep under sea level. This idea was used to help bolster the theory of plate tectonics.[*][7]

1.2.1 See also

- Hawaiian–Emperor seamount chain

- New England Seamount chain

- Kodiak–Bowie Seamount chain

- Evolution of Hawaiian volcanoes

- Hotspot (geology)

- Atoll

- Seamount

1.2.2 References

[1] Guyot *Encyclopædia Britannica Online*, 2010. Retrieved January 14, 2010.

[2] Seamount and guyot: Information and Much More from Answers.com

[3] Seamounts are made by extrusion of lavas piped upward in stages from sources within the Earth's mantle to vents on the seafloor. Seamounts provide data on movements of tectonic plates on which they ride, and on the rheology of the underlying lithosphere. The trend of a seamount chain traces the direction of motion of the lithospheric plate over a more or less fixed heat source in the underlying asthenosphere part of the Earth's mantle.

[4] Guyot

[5] Great Meteor Tablemount (volcanic mountain, Atlantic Ocean) - Britannica Online Encyclopedia

[6] |Sahfos||

[7] Bryson, Bill. A Short History of Nearly Everything. New York: Broadway, 2003. pg. 178

[8] http://etcweb.princeton.edu/CampusWWW/Companion/guyot_arnold.html Guyot, Arnold in *A Princeton Companion*

1.2.3 External links

- Wilde guyot map from Texas A&M

]

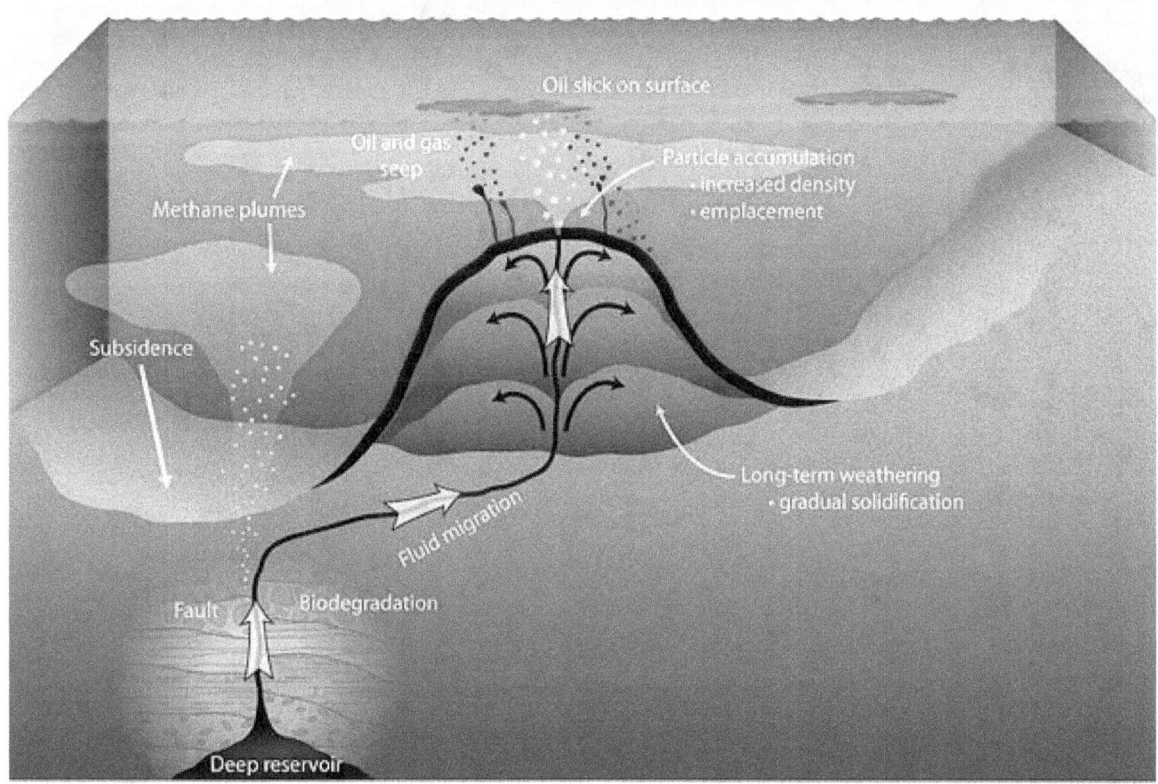

A diagram showing formation of an asphalt volcano and associated release of methane and oil.

1.3 Asphalt volcano

An **asphalt volcano** is a rare type of submarine volcano (seamount) first discovered in 2003. Several examples have been found: first, along the coasts of the United States and Mexico, and then in other regions of the world; a few are still active.[*][1] Resembling seamounts in structure, they are made entirely of asphalt, and form when natural oil seeps up from the Earth's crust underwater.

1.3.1 Formation and distribution

Asphalt volcanoes are ocean floor vents that erupt asphalt instead of lava. They were discovered in the Gulf of Mexico during an expedition of the research vessel SONNE, led by Gerhard Bohrmann of the DFG Research Center Ocean Margins. On these volcanoes a previously unknown highly diverse ecosystem at a water depth of 3,000 meters was discovered.[*][2]

The first asphalt volcanoes were discovered in 2003 by a research expedition to the Gulf of Mexico.[*][2] They are located on a seafloor hill named "Chapopote," Nahuatl for "tar." The site is located in a field of salt domes known as the Campeche Knolls, a series of steep hills formed from salt bodies that rise from underlying rock, a common feature in the gulf. The research team documented tar flows as wide as 20 m (66 ft) across. Also discovered alongside the asphalt were areas soaked with petroleum and methane hydrate, also spewed from the volcano. This kind of an environment proves attractive to chemical-loving bacteria and tubeworms, although the exact biogeochemical relationship is not yet known.[*][3]

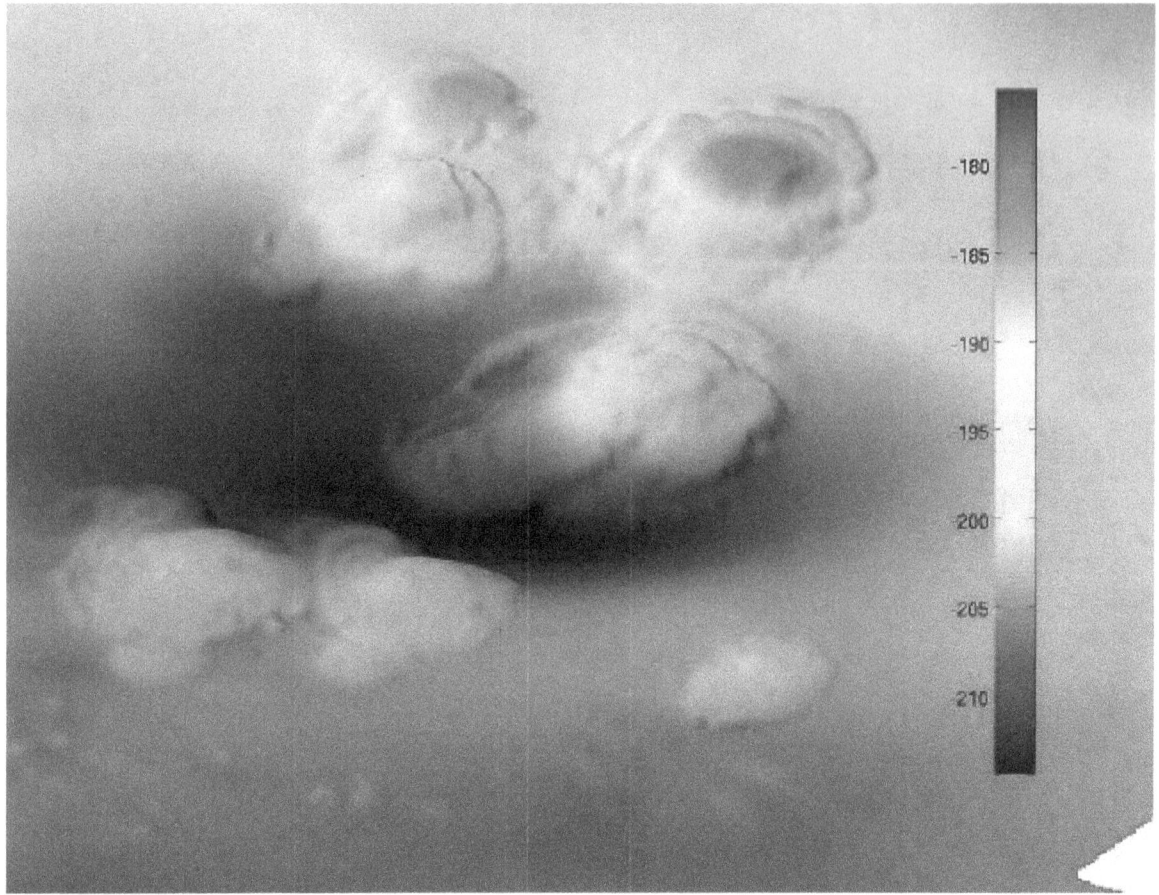

A bathymetric depiction of the seven asphalt volcanoes discovered west of Santa Barbara in 2007.

The tar is relatively hot when it comes out of the seafloor, but just like undersea lava flows, it is quickly cooled by the much colder seawater around it.*[2] This produces forms similar to the distinctive A'a and pahoehoe types of basalt lava flow seen in places like Hawaii. Another similarity is that the tar heats methane hydrate and causes it to explode into a free gas, similar to the action hot lava has on groundwater in phreatomagmatic eruptions.*[3]

The team proposed an asphalt volcano formation theory in a paper published in *Eos.**[2]*[4] The article suggested that water heated past the critical point underneath the seafloor found a passageway to the surface, most likely a salt dome, and carried with it a heavy load of hydrocarbons and dissolved minerals. A special property of such critically heated water is that it can mix with oils, whereas normal water cannot. The same process is attributed to the formation of black smokers. Once the water reaches the surface, it cools, and its carrying capacity drops.*[2] The lighter compounds in the mixture escape to the surface, while the tar and other heavier materials remain on the seafloor, eventually building up the asphalt volcano's structure.*[3]

In 2007, seven more such structures were discovered off the coast of Santa Barbara, California. The largest of these domes lies at a depth of 700 ft (213 m). The structures were larger than a football field and about as tall as a six-story building, all made completely out of asphalt. The unusual features were first noted by Ed Keller on bathymetric surveys conducted in the 1990s, and first viewed by a team led by David Valentine in 2007, utilizing *DSV Alvin*. Samples were brought up for testing at the university campus and the Woods Hole Oceanographic Institution.*[5]

Two further dives with DSV *Alvin* in 2009 and a detailed photographic survey of the area by the autonomous underwater vehicle *Sentry* showed many similarities to volcanic flows, including flow texture and cracking of the asphalt layers. Carbon dating puts the structures at between 30 and 40 thousand years old. They had at one time been a prolific source of methane. The two largest structures, less than 1 km (1 mi) apart, are pocked by pits and depressions, a sign of methane gas bubbling up long ago. Although the structures are still emitting residual gas, at present the amounts are too small to have any effect.[5] The amount of crude oil in the largest of the structures alone is "enough to fuel my Honda Civic for about half a billion miles. [However] the quality of the material is very poor...It's not worth something like light sweet crude," said Valentine. The petroleum in the structure is more viscous than that which is usually found in underground wells. This is because it has had less time to "bake" under the Earth's heat before being released. In addition, as much as 20% of its mass is made of "junk"—microscopic organisms, sand, and miscellaneous materials that gradually accumulated in the oil.[1]

Analysis of the samples collected from the mounds suggest that they required several decades, even centuries, to build up their current bulk, and that the volcanoes last erupted around 35,000 years ago. In addition they may account for a mysterious spike in oceanic methane concentrations around 35,000 years ago. Methane forms naturally alongside the petroleum underneath the structure, and while petroleum flows have long abated, some residual methane continues to bubble up.[2] This burst of methane would have caused a rapid increase in the population of methane-eating bacteria, which in turn caused a decrease in oxygen in the water, possibly causing a dead zone, in addition to the large amounts of crude oil released into the environment.[1]

Ecologically speaking, the presence of these structures provides a hard surface on which life can grow, as the surrounding ocean floor is generally muddy. This is similar to what happens on seamounts, resulting in their place as an ecological "hub." [1]

1.3.2   See also

- Petroleum seep

- Pockmark

- Mud volcano

1.3.3   References

[1] Than, Ker (April 26, 2010). "Huge Asphalt Volcanoes Discovered Off California". National Geographic. Archived from the original on 29 April 2010. Retrieved 30 April 2010.

[2] Asphalt volcanoes discovered Press Release 13. University of Bremen center for marine environmental sciences. May, 2004. Retrieved 5 July 2010.

[3] Alden, Andrew. "Asphalt Volcanism". about.com. Retrieved 29 April 2010.

[4] Hovland, M.; MacDonald I.R.; Rueslåtten H.; Johnsen H.K.; Naehr T.; Bohrmann G. (2005). "Chapopote Asphalt Volcano May Have Been Generated by Supercritical Water" (PDF). *EOS* **86** (42): 397–402. Bibcode:2005EOSTr..86..397H. doi:10.1029/2005EO420002. Retrieved 2010-04-30.

[5] Christopher Farwell, Sarah C. Bagby, Brian A. Clark, and Morgan Soloway, Robert K. Nelson, Dana Yoerger, and Richard Camilli, Tessa M. Hill, Oscar Pizarro, Christopher N. Roman (April 25, 2010). "Scientists Discover Underwater Asphalt Volcanoes". *Press Release 10-065*. National Science Foundation. Retrieved 30 April 2010.

1.4   Volcano

This article is about the geological feature. For other uses, see Volcano (disambiguation) and Volcanic (disambiguation).

A **volcano** is a rupture on the crust of a planetary-mass object, such as Earth, that allows hot lava, volcanic ash, and gases to escape from a magma chamber below the surface.

Earth's volcanoes occur because its crust is broken into 17 major, rigid tectonic plates that float on a hotter, softer layer in its mantle.[1] Therefore, on Earth, volcanoes are generally found where tectonic plates are diverging or converging. For example, a mid-oceanic ridge, such as the Mid-Atlantic Ridge, has volcanoes caused by divergent tectonic plates pulling apart; the Pacific Ring of Fire has volcanoes caused by convergent tectonic plates coming together. Volcanoes

Cleveland Volcano in the Aleutian Islands of Alaska photographed from the International Space Station, May 2006

can also form where there is stretching and thinning of the crust's interior plates, e.g., in the East African Rift and the Wells Gray-Clearwater volcanic field and Rio Grande Rift in North America. This type of volcanism falls under the umbrella of "plate hypothesis" volcanism.[2] Volcanism away from plate boundaries has also been explained as mantle plumes. These so-called "hotspots", for example Hawaii, are postulated to arise from upwelling diapirs with magma from the core–mantle boundary, 3,000 km deep in the Earth. Volcanoes are usually not created where two tectonic plates slide past one another.

Erupting volcanoes can pose many hazards, not only in the immediate vicinity of the eruption. One such hazard is that volcanic ash can be a threat to aircraft, in particular those with jet engines where ash particles can be melted by the high operating temperature; the melted particles then adhere to the turbine blades and alter their shape, disrupting the operation of the turbine. Large eruptions can affect temperature as ash and droplets of sulfuric acid obscure the sun and cool the Earth's lower atmosphere (or troposphere); however, they also absorb heat radiated up from the Earth, thereby warming the upper atmosphere (or stratosphere). Historically, so-called volcanic winters have caused catastrophic famines.

1.4.1   Etymology

The word *volcano* is derived from the name of Vulcano, a volcanic island in the Aeolian Islands of Italy whose name in turn originates from Vulcan, the name of a god of fire in Roman mythology.[3] The study of volcanoes is called volcanology, sometimes spelled *vulcanology*.

1.4.2   Plate tectonics

Main article: Plate tectonics

Ash plumes reached a height of 19 kilometres (12 mi) during the climactic explosive eruption at Mount Pinatubo, Philippines in 1991.

Divergent plate boundaries

Main article: Divergent boundary

At the mid-oceanic ridges, two tectonic plates diverge from one another as new oceanic crust is formed by the cooling and solidifying of hot molten rock. Because the crust is very thin at these ridges due to the pull of the tectonic plates, the release of pressure leads to adiabatic expansion and the partial melting of the mantle, causing volcanism and creating new oceanic crust. Most divergent plate boundaries are at the bottom of the oceans; therefore, most volcanic activity is submarine, forming new seafloor. Black smokers (also known as deep sea vents) are an example of this kind of volcanic activity. Where the mid-oceanic ridge is above sea-level, volcanic islands are formed, for example, Iceland.

Convergent plate boundaries

Main article: Convergent boundary

A 2007 eruptive column at Mount Etna producing volcanic ash, pumice and lava bombs

Subduction zones are places where two plates, usually an oceanic plate and a continental plate, collide. In this case, the oceanic plate subducts, or submerges under the continental plate forming a deep ocean trench just offshore. In a process called flux melting, water released from the subducting plate lowers the melting temperature of the overlying mantle wedge, creating magma. This magma tends to be very viscous due to its high silica content, so often does not reach the surface and cools at depth. When it does reach the surface, a volcano is formed. Typical examples of this kind of volcano are Mount Etna and the volcanoes in the Pacific Ring of Fire.

"Hotspots"

Main article: Hotspot (geology)

"Hotspots" is the name given to volcanic areas believed to be formed by mantle plumes, which are hypothesized to be columns of hot material rising from the core-mantle boundary in a fixed space that causes large-volume melting. Because tectonic plates move across them, each volcano becomes dormant and is eventually reformed as the plate advances over the postulated plume. The Hawaiian Islands have been suggested to have been formed in such a manner, as well as the Snake River Plain, with the Yellowstone Caldera being the part of the North American plate currently above the hot spot. This theory is currently under criticism, however.*[2]

1.4.3   Volcanic features

The most common perception of a volcano is of a conical mountain, spewing lava and poisonous gases from a crater at its summit; however, this describes just one of the many types of volcano. The features of volcanoes are much more complicated and their structure and behavior depends on a number of factors. Some volcanoes have rugged peaks formed by lava domes rather than a summit crater while others have landscape features such as massive plateaus. Vents that

Ubinas Volcano

issue volcanic material (including lava and ash) and gases (mainly steam and magmatic gases) can develop anywhere on the landform and may give rise to smaller cones such as Puʻu ʻŌʻō on a flank of Hawaii's Kīlauea. Other types of volcano include cryovolcanoes (or ice volcanoes), particularly on some moons of Jupiter, Saturn, and Neptune; and mud volcanoes, which are formations often not associated with known magmatic activity. Active mud volcanoes tend to involve temperatures much lower than those of igneous volcanoes except when the mud volcano is actually a vent of an igneous volcano.

Fissure vents

Main article: Fissure vent

Volcanic **fissure vents** are flat, linear fractures through which lava emerges.

Shield volcanoes

Main article: Shield volcano

Shield volcanoes, so named for their broad, shield-like profiles, are formed by the eruption of low-viscosity lava that can flow a great distance from a vent. They generally do not explode catastrophically. Since low-viscosity magma is typically low in silica, shield volcanoes are more common in oceanic than continental settings. The Hawaiian volcanic chain is a series of shield cones, and they are common in Iceland, as well.

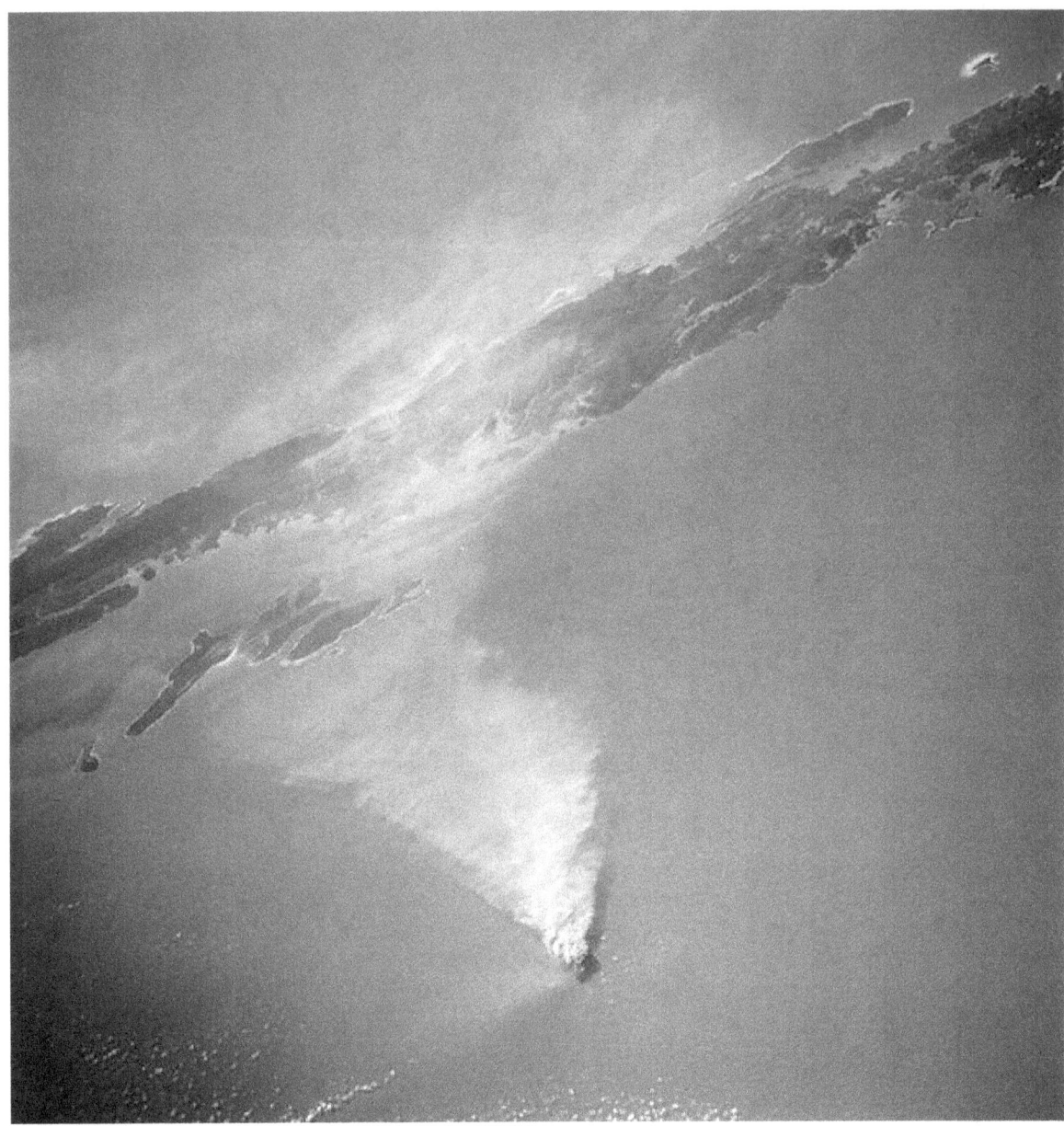

Aerial view of the Barren Island, Andaman Islands, India, during an eruption in 1995. It is the only active volcano in South Asia.

Lava domes

Main article: Lava dome

Lava domes are built by slow eruptions of highly viscous lava. They are sometimes formed within the crater of a previous volcanic eruption, as in the case of Mount Saint Helens, but can also form independently, as in the case of Lassen Peak. Like stratovolcanoes, they can produce violent, explosive eruptions, but their lava generally does not flow far from the originating vent.

Mount Shasta

Cryptodomes

Cryptodomes are formed when viscous lava is forced upward causing the surface to bulge. The 1980 eruption of Mount St. Helens was an example; lava beneath the surface of the mountain created an upward bulge which slid down the north side of the mountain.

Volcanic cones (cinder cones)

Main articles: volcanic cone and Cinder cone

 Volcanic cones or **cinder cones** result from eruptions of mostly small pieces of scoria and pyroclastics (both resemble cinders, hence the name of this volcano type) that build up around the vent. These can be relatively short-lived eruptions that produce a cone-shaped hill perhaps 30 to 400 meters high. Most cinder cones erupt only once. Cinder cones may form as flank vents on larger volcanoes, or occur on their own. Parícutin in Mexico and Sunset Crater in Arizona are examples of cinder cones. In New Mexico, Caja del Rio is a volcanic field of over 60 cinder cones.

Based on satellite images it was suggested that cinder cones might occur on other terrestrial bodies in the Solar system too; on the surface of Mars and the Moon.[*][4][*][5][*][6][*][7]

Stratovolcanoes (composite volcanoes)

Main article: Stratovolcano

Stratovolcanoes or **composite volcanoes** are tall conical mountains composed of lava flows and other ejecta in alternate layers, the strata that gives rise to the name. Stratovolcanoes are also known as composite volcanoes because they are created from multiple structures during different kinds of eruptions. Strato/composite volcanoes are made of cinders, ash,

Santa Ana Volcano, El Salvador. A close-up aerial view of the nested summit calderas and craters, along with the crater lake.

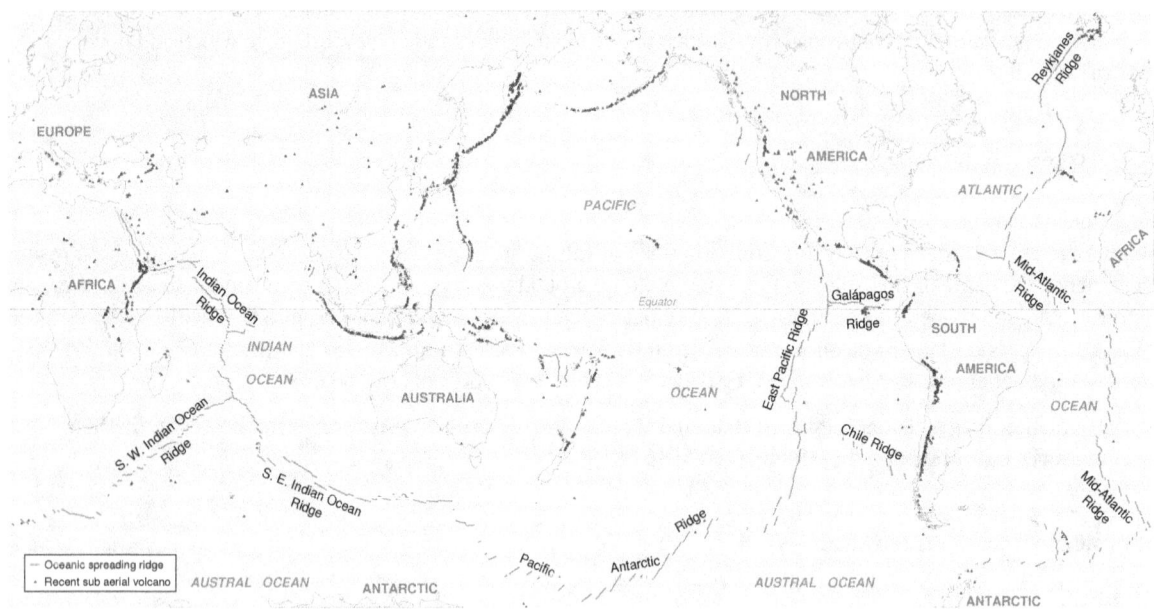

Map showing the divergent plate boundaries (OSR – Oceanic Spreading Ridges) and recent sub-aerial volcanoes

Lakagigar fissure vent in Iceland, source of the major world climate alteration of 1783–84

and lava. Cinders and ash pile on top of each other, lava flows on top of the ash, where it cools and hardens, and then the process repeats. Classic examples include Mt. Fuji in Japan, Mayon Volcano in the Philippines, and Mount Vesuvius and Stromboli in Italy.

Throughout recorded history, ash produced by the explosive eruption of stratovolcanoes has posed the greatest volcanic hazard to civilizations. Not only do stratovolcanoes have greater pressure build up from the underlying lava flow than shield volcanoes, but their fissure vents and monogenetic volcanic fields (volcanic cones) have more powerful eruptions, as they are many times under extension. They are also steeper than shield volcanoes, with slopes of 30–35° compared to slopes of generally 5–10°, and their loose tephra are material for dangerous lahars.*[8] Large pieces of tephra are called volcanic bombs. Big bombs can measure more than 4 feet(1.2 meters) across and weigh several tons.*[9]

Supervolcanoes

Main article: Supervolcano
See also: List of largest volcanic eruptions

A **supervolcano** usually has a large caldera and can produce devastation on an enormous, sometimes continental, scale. Such volcanoes are able to severely cool global temperatures for many years after the eruption due to the huge volumes of sulfur and ash released into the atmosphere. They are the most dangerous type of volcano. Examples include: Yellowstone Caldera in Yellowstone National Park and Valles Caldera in New Mexico (both western United States); Lake Taupo in New Zealand; Lake Toba in Sumatra, Indonesia; and Ngorongoro Crater in Tanzania. Because of the enormous area they may cover, supervolcanoes are hard to identify centuries after an eruption. Similarly, large igneous provinces are also

Skjaldbreiður, a shield volcano whose name means "broad shield"

considered supervolcanoes because of the vast amount of basalt lava erupted (even though the lava flow is non-explosive).

Submarine volcanoes

Main article: Submarine volcano
See also: Subaqueous volcano

Submarine volcanoes are common features of the ocean floor. In shallow water, active volcanoes disclose their presence by blasting steam and rocky debris high above the ocean's surface. In the ocean's deep, the tremendous weight of the water above prevents the explosive release of steam and gases; however, they can be detected by hydrophones and discoloration of water because of volcanic gases. Pillow lava is a common eruptive product of submarine volcanoes and is characterized by thick sequences of discontinuous pillow-shaped masses which form under water. Even large submarine eruptions may not disturb the ocean surface due to the rapid cooling effect and increased buoyancy of water (as compared to air) which often causes volcanic vents to form steep pillars on the ocean floor. Hydrothermal vents are common near these volcanoes, and some support peculiar ecosystems based on dissolved minerals. Over time, the formations created by submarine volcanoes may become so large that they break the ocean surface as new islands or floating pumice rafts.

Subglacial volcanoes

Main article: Subglacial volcano

Subglacial volcanoes develop underneath icecaps. They are made up of flat lava which flows at the top of extensive pillow lavas and palagonite. When the icecap melts, the lava on top collapses, leaving a flat-topped mountain. These volcanoes are also called table mountains, tuyas, or (uncommonly) mobergs. Very good examples of this type of volcano can be seen in Iceland, however, there are also tuyas in British Columbia. The origin of the term comes from Tuya Butte, which

Izalco (volcano), located in the Cordillera de Apaneca volcanic range complex in El Salvador. Only a few generations old, Izalco is the youngest and best known cone volcano. Izalco erupted almost continuously from 1770 (when it formed) to 1958, earning it the nickname of "Lighthouse of the Pacific".

is one of the several tuyas in the area of the Tuya River and Tuya Range in northern British Columbia. Tuya Butte was the first such landform analyzed and so its name has entered the geological literature for this kind of volcanic formation. The Tuya Mountains Provincial Park was recently established to protect this unusual landscape, which lies north of Tuya Lake and south of the Jennings River near the boundary with the Yukon Territory.

Mud volcanoes

Main article: Mud volcano

Mud volcanoes or **mud domes** are formations created by geo-excreted liquids and gases, although there are several processes which may cause such activity. The largest structures are 10 kilometers in diameter and reach 700 meters high.

1.4.4   Erupted material

Lava composition

Another way of classifying volcanoes is by the *composition of material erupted* (lava), since this affects the shape of the volcano. Lava can be broadly classified into 4 different compositions (Cas & Wright, 1987):

- If the erupted magma contains a high percentage (>63%) of silica, the lava is called felsic.

  - Felsic lavas (dacites or rhyolites) tend to be highly viscous (not very fluid) and are erupted as domes or short,

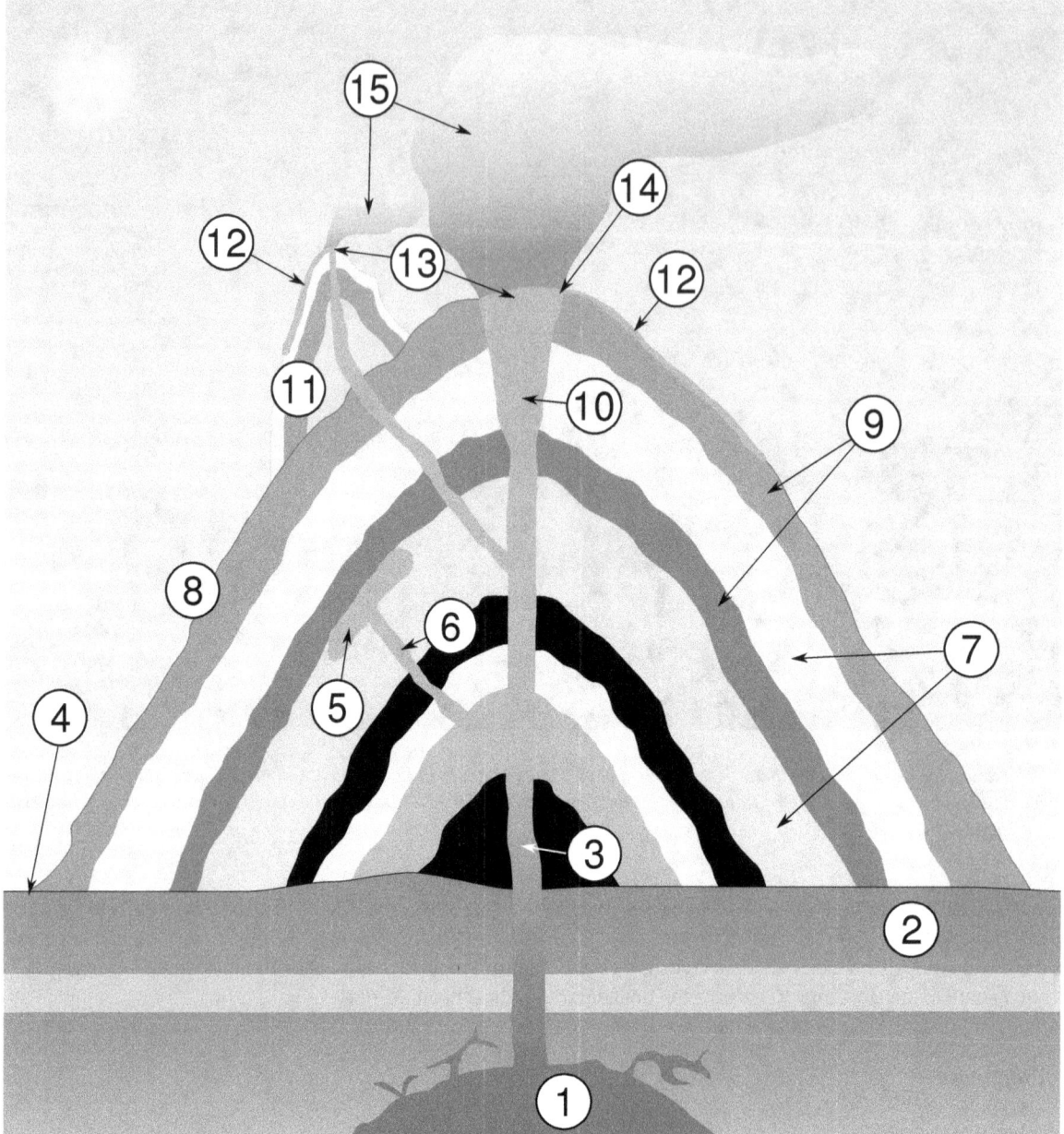

stubby flows. Viscous lavas tend to form stratovolcanoes or lava domes. Lassen Peak in California is an example of a volcano formed from felsic lava and is actually a large lava dome.

- Because siliceous magmas are so viscous, they tend to trap volatiles (gases) that are present, which cause the magma to erupt catastrophically, eventually forming stratovolcanoes. Pyroclastic flows (ignimbrites) are highly hazardous products of such volcanoes, since they are composed of molten volcanic ash too heavy to go up into the atmosphere, so they hug the volcano's slopes and travel far from their vents during large eruptions. Temperatures as high as 1,200 °C are known to occur in pyroclastic flows, which will incinerate everything flammable in their path and thick layers of hot pyroclastic flow deposits can be laid down, often up to many meters thick. Alaska's Valley of Ten Thousand Smokes, formed by the eruption of Novarupta near Katmai in 1912, is an example of a thick pyroclastic flow or ignimbrite deposit. Volcanic ash that is light enough to be erupted high into the Earth's atmosphere may travel many kilometres before it falls back to ground as a tuff.

- If the erupted magma contains 52–63% silica, the lava is of *intermediate* composition.

Pāhoehoe lava flow on Hawaii. The picture shows overflows of a main lava channel.

- These "andesitic" volcanoes generally only occur above subduction zones (e.g. Mount Merapi in Indonesia).
- Andesitic lava is typically formed at convergent boundary margins of tectonic plates, by several processes:
  - Hydration melting of peridotite and fractional crystallization
  - Melting of subducted slab containing sediments
  - Magma mixing between felsic rhyolitic and mafic basaltic magmas in an intermediate reservoir prior to emplacement or lava flow.
- If the erupted magma contains <52% and >45% silica, the lava is called mafic (because it contains higher percentages of magnesium (Mg) and iron (Fe)) or basaltic. These lavas are usually much less viscous than rhyolitic lavas, depending on their eruption temperature; they also tend to be hotter than felsic lavas. Mafic lavas occur in a wide range of settings:
  - At mid-ocean ridges, where two oceanic plates are pulling apart, basaltic lava erupts as pillows to fill the gap;
  - Shield volcanoes (e.g. the Hawaiian Islands, including Mauna Loa and Kilauea), on both oceanic and continental crust;
  - As continental flood basalts.
- Some erupted magmas contain <=45% silica and produce ultramafic lava. Ultramafic flows, also known as komatiites, are very rare; indeed, very few have been erupted at the Earth's surface since the Proterozoic, when the planet's heat flow was higher. They are (or were) the hottest lavas, and probably more fluid than common mafic lavas.

Lava texture

Two types of lava are named according to the surface texture: ʻAʻa (pronounced [ˈʔaʔa]) and pāhoehoe ([paːˈhoeˈhoe]), both Hawaiian words. ʻAʻa is characterized by a rough, clinkery surface and is the typical texture of viscous lava flows.

The Stromboli stratovolcano off the coast of Sicily has erupted continuously for thousands of years, giving rise to the term strombolian eruption.

However, even basaltic or mafic flows can be erupted as 'a'a flows, particularly if the eruption rate is high and the slope is steep.

Pāhoehoe is characterized by its smooth and often ropey or wrinkly surface and is generally formed from more fluid lava flows. Usually, only mafic flows will erupt as pāhoehoe, since they often erupt at higher temperatures or have the proper chemical make-up to allow them to flow with greater fluidity.

1.4.5 Volcanic activity

Popular classification of volcanoes

A popular way of classifying magmatic volcanoes is by their frequency of eruption, with those that erupt regularly called **active**, those that have erupted in historical times but are now quiet called **dormant** or **inactive**, and those that have not erupted in historical times called **extinct**. However, these popular classifications—extinct in particular—are practically meaningless to scientists. They use classifications which refer to a particular volcano's formative and eruptive processes and resulting shapes, which was explained above.

Active There is no consensus among volcanologists on how to define an "active" volcano. The lifespan of a volcano can vary from months to several million years, making such a distinction sometimes meaningless when compared to the lifespans of humans or even civilizations. For example, many of Earth's volcanoes have erupted dozens of times in the past few thousand years but are not currently showing signs of eruption. Given the long lifespan of such volcanoes, they are very active. By human lifespans, however, they are not.

Scientists usually consider a volcano to be *erupting* or *likely to erupt* if it is currently erupting, or showing signs of unrest such as unusual earthquake activity or significant new gas emissions. Most scientists consider a volcano *active* if it has

San Miguel (volcano), El Salvador. On December 29, 2013, San Miguel volcano, also known as "Chaparrastique", erupted at 10:30 local time, spewing a large column of ash and smoke into the sky; the eruption, the first in 11 years, was seen from space and prompted the evacuation of thousands of people living in a 3 km radius around the volcano.

erupted in the last 10,000 years (Holocene times) – the Smithsonian Global Volcanism Program uses this definition of *active*. Most volcanoes are situated on the Pacific Ring of Fire.*[10] An estimated 500 million people live near active volcanoes.*[10]

Historical time (or recorded history) is another timeframe for *active*.*[11]*[12] The *Catalogue of the Active Volcanoes of the World*, published by the International Association of Volcanology, uses this definition, by which there are more than 500 active volcanoes.*[11] However, the span of recorded history differs from region to region. In China and the Mediterranean, it reaches back nearly 3,000 years, but in the Pacific Northwest of the United States and Canada, it reaches back less than 300 years, and in Hawaii and New Zealand, only around 200 years.*[11]

As of 2013, the following are considered Earth's most active volcanoes:*[13]

- Kīlauea, the famous Hawaiian volcano, has been in continuous, effusive eruption since 1983, and has the longest-observed lava lake.

- Mount Etna and nearby Stromboli, two Mediterranean volcanoes in "almost continuous eruption" since antiquity.

- Mount Yasur, in Vanuatu, has been erupting "nearly continuously" for over 800 years.

The longest currently ongoing (but not necessarily continuous) volcanic eruptive phases are:*[14]

- Mount Yasur, 111 years

- Mount Etna, 109 years

- Stromboli, 108 years

Ash plume from San Miguel (volcano) "Chaparrastique", seen from a satellite, as it heads towards the Pacific Ocean from the El Salvador Central America coast, December 29, 2013

- Santa María, 101 years

- Sangay, 94 years

Other very active volcanoes include:

- Mount Nyiragongo and its neighbor, Nyamuragira, are Africa's most active volcanoes .

- Piton de la Fournaise, in Réunion, erupts frequently enough to be a tourist attraction.

- Erta Ale, in the Afar Triangle, has maintained a lava lake since at least 1906.

- Mount Erebus, in Antarctica, has maintained a lava lake since at least 1972.

- Mount Merapi

- Whakaari / White Island, has been in continuous state of smoking since its discovery in 1769.

- Ol Doinyo Lengai

- Ambrym

- Arenal Volcano

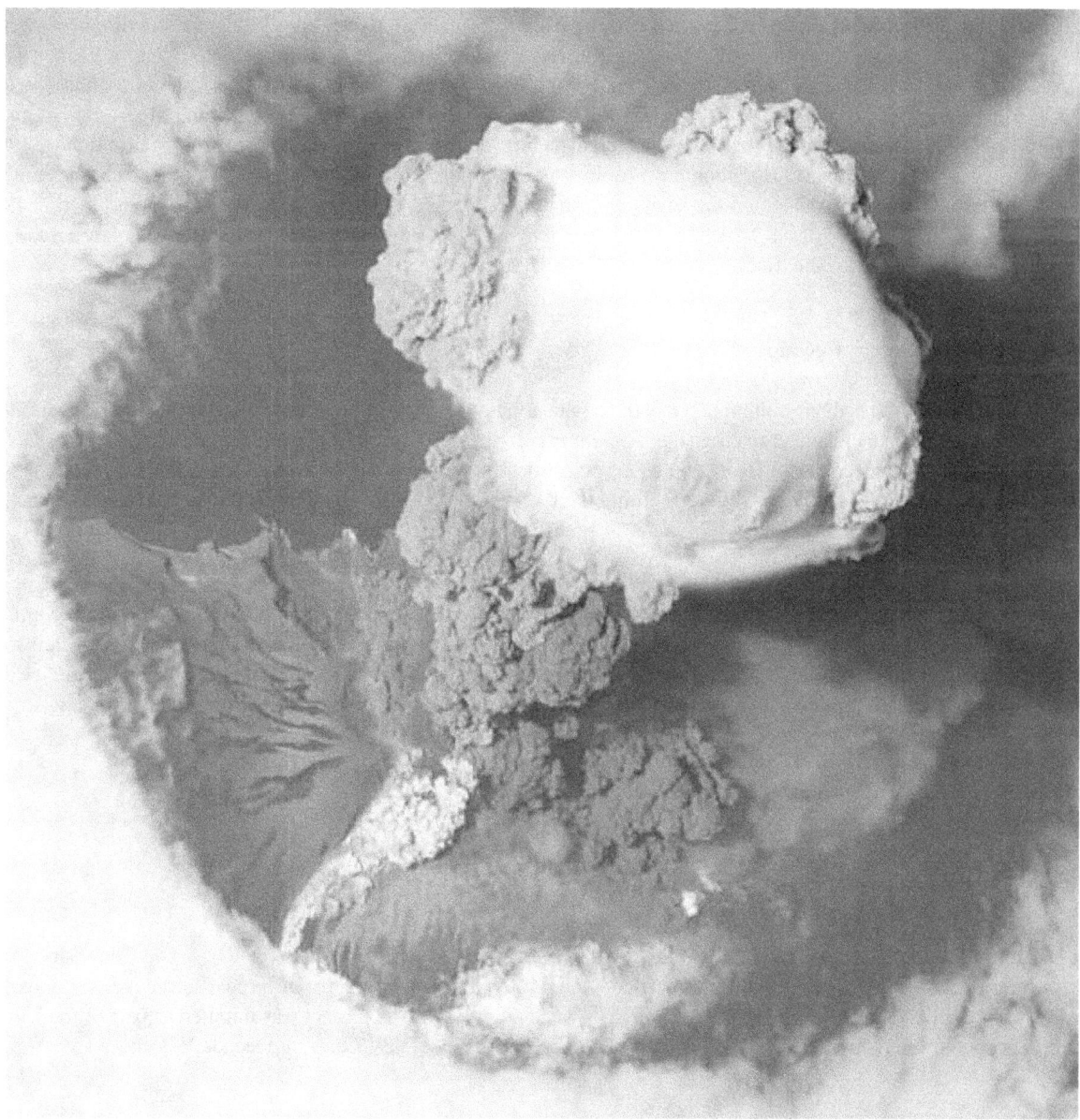

Sarychev Peak eruption, Matua Island, oblique satellite view

- Pacaya

- Klyuchevskaya Sopka

- Sheveluch

Extinct   Extinct volcanoes are those that scientists consider unlikely to erupt again, because the volcano no longer has a magma supply. Examples of extinct volcanoes are many volcanoes on the Hawaiian – Emperor seamount chain in the Pacific Ocean, Hohentwiel, Shiprock and the Zuidwal volcano in the Netherlands. Edinburgh Castle in Scotland is famously located atop an extinct volcano. Otherwise, whether a volcano is truly extinct is often difficult to determine. Since "supervolcano" calderas can have eruptive lifespans sometimes measured in millions of years, a caldera that has not produced an eruption in tens of thousands of years is likely to be considered dormant instead of extinct. Some volcanologists refer to extinct volcanoes as inactive, though the term is now more commonly used for dormant volcanoes once thought to be extinct.

Dormant It is difficult to distinguish an extinct volcano from a dormant (inactive) one. Volcanoes are often considered to be extinct if there are no written records of its activity. Nevertheless, volcanoes may remain dormant for a long period of time. For example, Yellowstone has a repose/recharge period of around 700,000 years, and Toba of around 380,000 years.[15] Vesuvius was described by Roman writers as having been covered with gardens and vineyards before its eruption of AD 79, which destroyed the towns of Herculaneum and Pompeii. Before its catastrophic eruption of 1991, Pinatubo was an inconspicuous volcano, unknown to most people in the surrounding areas. Two other examples are the long-dormant Soufrière Hills volcano on the island of Montserrat, thought to be extinct before activity resumed in 1995 and Fourpeaked Mountain in Alaska, which, before its September 2006 eruption, had not erupted since before 8000 BC and had long been thought to be extinct.

Technical classification of volcanoes

Volcanic-alert level The three common popular classifications of volcanoes can be subjective and some volcanoes thought to have been extinct have erupted again. To help prevent people from falsely believing they are not at risk when living on or near a volcano, countries have adopted new classifications to describe the various levels and stages of volcanic activity.[16] Some alert systems use different numbers or colors to designate the different stages. Other systems use colors and words. Some systems use a combination of both.

Volcano warning schemes of the United States The United States Geological Survey (USGS) has adopted a common system nationwide for characterizing the level of unrest and eruptive activity at volcanoes. The new volcano alert-level system classifies volcanoes now as being in a normal, advisory, watch or warning stage. Additionally, colors are used to denote the amount of ash produced. Details of the US system can be found at Volcano warning schemes of the United States.

1.4.6 Decade volcanoes

Main articles: Lists of volcanoes and Decade Volcanoes

The Decade Volcanoes are 17 volcanoes identified by the International Association of Volcanology and Chemistry of the Earth's Interior (IAVCEI) as being worthy of particular study in light of their history of large, destructive eruptions and proximity to populated areas. They are named Decade Volcanoes because the project was initiated as part of the United Nations-sponsored International Decade for Natural Disaster Reduction. The 17 current Decade Volcanoes are

1.4.7 Effects of volcanoes

There are many different types of volcanic eruptions and associated activity: phreatic eruptions (steam-generated eruptions), explosive eruption of high-silica lava (e.g., rhyolite), effusive eruption of low-silica lava (e.g., basalt), pyroclastic flows, lahars (debris flow) and carbon dioxide emission. All of these activities can pose a hazard to humans. Earthquakes, hot springs, fumaroles, mud pots and geysers often accompany volcanic activity.

Volcanic gases

The concentrations of different volcanic gases can vary considerably from one volcano to the next. Water vapor is typically the most abundant volcanic gas, followed by carbon dioxide[17] and sulfur dioxide. Other principal volcanic gases include hydrogen sulfide, hydrogen chloride, and hydrogen fluoride. A large number of minor and trace gases are also found in volcanic emissions, for example hydrogen, carbon monoxide, halocarbons, organic compounds, and volatile metal chlorides.

Large, explosive volcanic eruptions inject water vapor (H_2O), carbon dioxide (CO_2), sulfur dioxide (SO_2), hydrogen chloride (HCl), hydrogen fluoride (HF) and ash (pulverized rock and pumice) into the stratosphere to heights of 16–32 kilometres (10–20 mi) above the Earth's surface. The most significant impacts from these injections come from the conversion of sulfur dioxide to sulfuric acid (H_2SO_4), which condenses rapidly in the stratosphere to form fine sulfate aerosols. It is worth mentioning that the SO_2 emissions alone of two different eruptions are sufficient to compare their potential climatic impact.[18] The aerosols increase the Earth's albedo—its reflection of radiation from the Sun back into space – and thus cool the Earth's lower atmosphere or troposphere; however, they also absorb heat radiated up from the Earth, thereby warming the stratosphere. Several eruptions during the past century have caused a decline in the average temperature at the Earth's surface of up to half a degree (Fahrenheit scale) for periods of one to three years – sulfur dioxide from the eruption of Huaynaputina probably caused the Russian famine of 1601–1603.[19]

Significant consequences

One proposed volcanic winter happened c. 70,000 years ago following the supereruption of Lake Toba on Sumatra island in Indonesia.[20] According to the Toba catastrophe theory to which some anthropologists and archeologists subscribe, it had global consequences,[21] killing most humans then alive and creating a population bottleneck that affected the genetic inheritance of all humans today.[22] The 1815 eruption of Mount Tambora created global climate anomalies that became known as the "Year Without a Summer" because of the effect on North American and European weather.[23] Agricultural crops failed and livestock died in much of the Northern Hemisphere, resulting in one of the worst famines of the 19th century.[24] The freezing winter of 1740–41, which led to widespread famine in northern Europe, may also owe its origins to a volcanic eruption.[25]

It has been suggested that volcanic activity caused or contributed to the End-Ordovician, Permian-Triassic, Late Devonian mass extinctions, and possibly others. The massive eruptive event which formed the Siberian Traps, one of the largest known volcanic events of the last 500 million years of Earth's geological history, continued for a million years and is considered to be the likely cause of the "Great Dying" about 250 million years ago,[26] which is estimated to have killed 90% of species existing at the time.[27]

Acid rain

The sulfate aerosols also promote complex chemical reactions on their surfaces that alter chlorine and nitrogen chemical species in the stratosphere. This effect, together with increased stratospheric chlorine levels from chlorofluorocarbon pollution, generates chlorine monoxide (ClO), which destroys ozone (O_3). As the aerosols grow and coagulate, they settle down into the upper troposphere where they serve as nuclei for cirrus clouds and further modify the Earth's radiation balance. Most of the hydrogen chloride (HCl) and hydrogen fluoride (HF) are dissolved in water droplets in the eruption cloud and quickly fall to the ground as acid rain. The injected ash also falls rapidly from the stratosphere; most of it is removed within several days to a few weeks. Finally, explosive volcanic eruptions release the greenhouse gas carbon dioxide and thus provide a deep source of carbon for biogeochemical cycles.[28]

Gas emissions from volcanoes are a natural contributor to acid rain. Volcanic activity releases about 130 to 230 teragrams (145 million to 255 million short tons) of carbon dioxide each year.[29] Volcanic eruptions may inject aerosols into the Earth's atmosphere. Large injections may cause visual effects such as unusually colorful sunsets and affect global climate mainly by cooling it. Volcanic eruptions also provide the benefit of adding nutrients to soil through the weathering process of volcanic rocks. These fertile soils assist the growth of plants and various crops. Volcanic eruptions can also create new islands, as the magma cools and solidifies upon contact with the water.

Hazards

Ash thrown into the air by eruptions can present a hazard to aircraft, especially jet aircraft where the particles can be melted by the high operating temperature; the melted particles then adhere to the turbine blades and alter their shape, disrupting the operation of the turbine. Dangerous encounters in 1982 after the eruption of Galunggung in Indonesia, and 1989 after the eruption of Mount Redoubt in Alaska raised awareness of this phenomenon. Nine Volcanic Ash Advisory Centers were established by the International Civil Aviation Organization to monitor ash clouds and advise pilots accordingly. The 2010 eruptions of Eyjafjallajökull caused major disruptions to air travel in Europe.

1.4.8 Volcanoes on other planetary bodies

See also: List of extraterrestrial volcanoes, Geology of the Moon, Volcanology of Mars, Volcanology of Io and Volcanology of Venus

The Earth's Moon has no large volcanoes and no current volcanic activity, although recent evidence suggests it may still possess a partially molten core.[*][30] However, the Moon does have many volcanic features such as maria (the darker patches seen on the moon), rilles and domes.

The planet Venus has a surface that is 90% basalt, indicating that volcanism played a major role in shaping its surface. The planet may have had a major global resurfacing event about 500 million years ago,[*][31] from what scientists can tell from the density of impact craters on the surface. Lava flows are widespread and forms of volcanism not present on Earth occur as well. Changes in the planet's atmosphere and observations of lightning have been attributed to ongoing volcanic eruptions, although there is no confirmation of whether or not Venus is still volcanically active. However, radar sounding by the Magellan probe revealed evidence for comparatively recent volcanic activity at Venus's highest volcano Maat Mons, in the form of ash flows near the summit and on the northern flank.

There are several extinct volcanoes on Mars, four of which are vast shield volcanoes far bigger than any on Earth. They include Arsia Mons, Ascraeus Mons, Hecates Tholus, Olympus Mons, and Pavonis Mons. These volcanoes have been extinct for many millions of years,[*][32] but the European *Mars Express* spacecraft has found evidence that volcanic activity may have occurred on Mars in the recent past as well.[*][32]

Jupiter's moon Io is the most volcanically active object in the solar system because of tidal interaction with Jupiter. It is covered with volcanoes that erupt sulfur, sulfur dioxide and silicate rock, and as a result, Io is constantly being resurfaced. Its lavas are the hottest known anywhere in the solar system, with temperatures exceeding 1,800 K (1,500 °C). In February 2001, the largest recorded volcanic eruptions in the solar system occurred on Io.[*][33] Europa, the smallest of Jupiter's Galilean moons, also appears to have an active volcanic system, except that its volcanic activity is entirely in the form of water, which freezes into ice on the frigid surface. This process is known as cryovolcanism, and is apparently most common on the moons of the outer planets of the solar system.

In 1989 the Voyager 2 spacecraft observed cryovolcanoes (ice volcanoes) on Triton, a moon of Neptune, and in 2005 the Cassini–Huygens probe photographed fountains of frozen particles erupting from Enceladus, a moon of Saturn.[*][34][*][35] The ejecta may be composed of water, liquid nitrogen, dust, or methane compounds. Cassini–Huygens also found evidence of a methane-spewing cryovolcano on the Saturnian moon Titan, which is believed to be a significant source of the methane found in its atmosphere.[*][36] It is theorized that cryovolcanism may also be present on the Kuiper Belt Object Quaoar.

A 2010 study of the exoplanet COROT-7b, which was detected by transit in 2009, studied that tidal heating from the host star very close to the planet and neighboring planets could generate intense volcanic activity similar to Io.[*][37]

1.4.9 Traditional beliefs about volcanoes

Many ancient accounts ascribe volcanic eruptions to supernatural causes, such as the actions of gods or demigods. To the ancient Greeks, volcanoes' capricious power could only be explained as acts of the gods, while 16th/17th-century German astronomer Johannes Kepler believed they were ducts for the Earth's tears.[*][38] One early idea counter to this was proposed by Jesuit Athanasius Kircher (1602–1680), who witnessed eruptions of Mount Etna and Stromboli, then visited the crater of Vesuvius and published his view of an Earth with a central fire connected to numerous others caused by the burning of sulfur, bitumen and coal.

Various explanations were proposed for volcano behavior before the modern understanding of the Earth's mantle structure as a semisolid material was developed. For decades after awareness that compression and radioactive materials may be heat sources, their contributions were specifically discounted. Volcanic action was often attributed to chemical reactions and a thin layer of molten rock near the surface.

1.4.10   See also

- Global Volcanism Program

- List of extraterrestrial volcanoes

- Maritime impacts of volcanic eruptions

- Prediction of volcanic activity

- Timetable of major worldwide volcanic eruptions

- Volcanic Explosivity Index

- Volcano Number

- Volcano observatory

1.4.11   References

[1] NSTA Press / Archive.Org (2007). "Earthquakes, Volcanoes, and Tsunamis" (PDF). *Resources for Environmental Literacy.* Retrieved April 22, 2014.

[2] Foulger, G.R. (2010). *Plates vs. Plumes: A Geological Controversy.* Wiley-Blackwell. ISBN 978-1-4051-6148-0.

[3] Douglas Harper (November 2001). "Volcano". *Online Etymology Dictionary.* Retrieved June 11, 2009.

[4] Wood, C. A., 1979b. Cinder cones on Earth, Moon and Mars. Lunar Planet. Sci. X, 1370–1372.

[5] Meresse, S.; Costard, F. O.; Mangold, N.; Masson, P.; Neukum, G. (2008). "Formation and evolution of the chaotic terrains by subsidence and magmatism: Hydraotes Chaos, Mars". *Icarus* **194** (2): 487. doi:10.1016/j.icarus.2007.10.023.

[6] Brož, P.; Hauber, E. (2012). "A unique volcanic field in Tharsis, Mars: Pyroclastic cones as evidence for explosive eruptions". *Icarus* **218**: 88. doi:10.1016/j.icarus.2011.11.030.

[7] Lawrence, S. J.; Stopar, J. D.; Hawke, B. R.; Greenhagen, B. T.; Cahill, J. T. S.; Bandfield, J. L.; Jolliff, B. L.; Denevi, B. W.; Robinson, M. S.; Glotch, T. D.; Bussey, D. B. J.; Spudis, P. D.; Giguere, T. A.; Garry, W. B. (2013). "LRO observations of morphology and surface roughness of volcanic cones and lobate lava flows in the Marius Hills". *Journal of Geophysical Research: Planets* **118** (4): 615. doi:10.1002/jgre.20060.

[8] Lockwood, John P.; Hazlett, Richard W. (2010). *Volcanoes: Global Perspectives.* p. 552. ISBN 978-1-4051-6250-0.

[9] Berger, Melvin, Gilda Berger, and Higgins Bond. "Volcanoes-why and how." Why do volcanoes blow their tops?: Questions and answers about volcanoes and earthquakes. New York: Scholastic, 1999. 7. Print.

[10] "Volcanoes". European Space Agency. 2009. Retrieved August 16, 2012.

[11] Decker, Robert Wayne; Decker, Barbara (1991). *Mountains of Fire: The Nature of Volcanoes.* Cambridge University Press. p. 7. ISBN 0-521-31290-6. Retrieved August 16, 2012.

[12] Tilling, Robert I. (1997). "Volcano environments". *Volcanoes.* Denver, Colorado: U.S. Department of the Interior, U.S. Geological Survey. Retrieved August 16, 2012. There are more than 500 active volcanoes (those that have erupted at least once within recorded history) in the world

[13] "The most active volcanoes in the world". VolcanoDiscovery.com. Retrieved 3 August 2013.

[14] "The World's Five Most Active Volcanoes". livescience.com. Retrieved 4 August 2013.

[15] Chesner, C.A.; Rose, J.A.; Deino, W.I.; Drake, R.; Westgate, A. (March 1991). "Eruptive History of Earth's Largest Quaternary caldera (Toba, Indonesia) Clarified" (PDF). *Geology* **19** (3): 200–203. Bibcode:1991Geo....19..200C. doi:10.1130/0091-7613(1991)019<0200:EHOESL>2.3.CO;2. Retrieved January 20, 2010.

[16] "Volcanic Alert Levels of Various Countries". Volcanolive.com. Retrieved August 22, 2011.

[17] Pedone, M.; Aiuppa, A.; Giudice, G.; Grassa, F.; Francofonte, V.; Bergsson, B.; Ilyinskaya, E. (2014). "Tunable diode laser measurements of hydrothermal/volcanic CO_2 and implications for the global CO_2 budget." (PDF). *Solid Earth* **5**: 1209–1221. doi:10.5194/se-5-1209-2014.

[18] Miles, M. G.; Grainger, R. G.; Highwood, E. J. (2004). "The significance of volcanic eruption strength and frequency for climate" (PDF). *Quarterly Journal of the Royal Meteorological Society* **130**: 2361–2376. doi:10.1256/qj.30.60.

[19] University of California – Davis (April 25, 2008). "Volcanic Eruption Of 1600 Caused Global Disruption". *ScienceDaily.*

[20] "Supervolcano Eruption – In Sumatra – Deforested India 73,000 Years Ago". *ScienceDaily.* November 24, 2009.

[21] "The new batch – 150,000 years ago". BBC – Science & Nature – The evolution of man.

[22] "When humans faced extinction". BBC. June 9, 2003. Retrieved January 5, 2007.

[23] *Volcanoes in human history: the far-reaching effects of major eruptions.* Jelle Zeilinga de Boer, Donald Theodore Sanders (2002). Princeton University Press. p. 155. ISBN 0-691-05081-3

[24] Oppenheimer, Clive (2003). "Climatic, environmental and human consequences of the largest known historic eruption: Tambora volcano (Indonesia) 1815". *Progress in Physical Geography* **27** (2): 230–259. doi:10.1191/0309133303pp379ra.

[25] "Ó Gráda, C.: Famine: A Short History". Princeton University Press.

[26] "Yellowstone's Super Sister". Discovery Channel.

[27] Benton M J (2005). *When Life Nearly Died: The Greatest Mass Extinction of All Time.* Thames & Hudson. ISBN 978-0-500-28573-2.

[28] McGee, Kenneth A.; Doukas, Michael P.; Kessler, Richard; Gerlach, Terrence M. (May 1997). "Impacts of Volcanic Gases on Climate, the Environment, and People". United States Geological Survey. Retrieved 9 August 2014. *This article incorporates text from this source, which is in the public domain.*

[29] "Volcanic Gases and Their Effects". U.S. Geological Survey. Retrieved June 16, 2007.

[30] M. A. Wieczorek, B. L. Jolliff, A. Khan, M. E. Pritchard, B. P. Weiss, J. G. Williams, L. L. Hood, K. Righter, C. R. Neal, C. K. Shearer, I. S. McCallum, S. Tompkins, B. R. Hawke, C. Peterson, J, J. Gillis, B. Bussey (2006). "The Constitution and Structure of the Lunar Interior". *Reviews in Mineralogy and Geochemistry* **60** (1): 221–364. doi:10.2138/rmg.2006.60.3.

[31] Bindschadler, D. L. (1995). "Magellan: A new view of Venus' geology and geophysics". *Reviews of Geophysics* **33**: 459. doi:10.1029/95RG00281. Retrieved 28 September 2015.

[32] "Glacial, volcanic and fluvial activity on Mars: latest images". European Space Agency. February 25, 2005. Retrieved August 17, 2006.

[33] "Exceptionally bright eruption on Io rivals largest in Solar System", November 13, 2002.

[34] "Cassini Finds an Atmosphere on Saturn's Moon Enceladus". *PPARC*. 16 March 2005. Archived from the original on 2007-03-10. Retrieved 4 July 2014.

[35] Smith, Yvette (March 15, 2012). "Enceladus, Saturn's Moon". *Image of the Day Gallery*. NASA. Retrieved 4 July 2014.

[36] "Hydrocarbon volcano discovered on Titan". Newscientist.com. June 8, 2005. Retrieved October 24, 2010.

[37] Jaggard, Victoria (February 5, 2010). ""Super Earth" May Really Be New Planet Type: Super-Io". *National Geographic web site daily news.* National Geographic Society. Retrieved March 11, 2010.

[38] Williams, Micheal (November 2007). "Hearts of fire". *Morning Calm* (Korean Air Lines) (11–2007): 6.

1.4.12   Further reading

- Cas, R.A.F. and J.V. Wright, 1987. Volcanic Successions. Unwin Hyman Inc. 528p. ISBN 0-04-552022-4

- Macdonald, Gordon and Agatin T. Abbott. (1970). Volcanoes in the Sea. University of Hawaii Press, Honolulu. 441 p.

- Marti, Joan and Ernst, Gerald. (2005). *Volcanoes and the Environment.* Cambridge University Press. ISBN 0-521-59254-2.

- Ollier, Cliff. (1988). Volcanoes. Basil Blackwell, Oxford, UK, ISBN 0-631-15664-X (hardback), ISBN 0-631-15977-0 (paperback).

- Sigurðsson, Haraldur, ed. (1999). *Encyclopedia of Volcanoes.* Academic Press. ISBN 0-12-643140-X. This is a reference aimed at geologists, but many articles are accessible to non-professionals.

1.4.13   External links

- Volcanoes at DMOZ

- Volcano, U.S. Federal Emergency Management Agency FEMA

- Volcano World

- Volcanos (Worsley School)

Fresco with Mount Vesuvius behind Bacchus and Agathodaemon, as seen in Pompeii's House of the Centenary

Kīlauea lava entering the sea.

Lava flows at Holuhraun, Iceland, September 2014

Nyiragongo's lava lake

Fourpeaked volcano, Alaska, in September 2006 after being thought extinct for over 10,000 years

Mount Rinjani eruption in 1994, in Lombok, Indonesia

Narcondam Island, India, is classified as a dormant volcano by the Geological Survey of India

Koryaksky volcano towering over Petropavlovsk-Kamchatsky on Kamchatka Peninsula, Far Eastern Russia

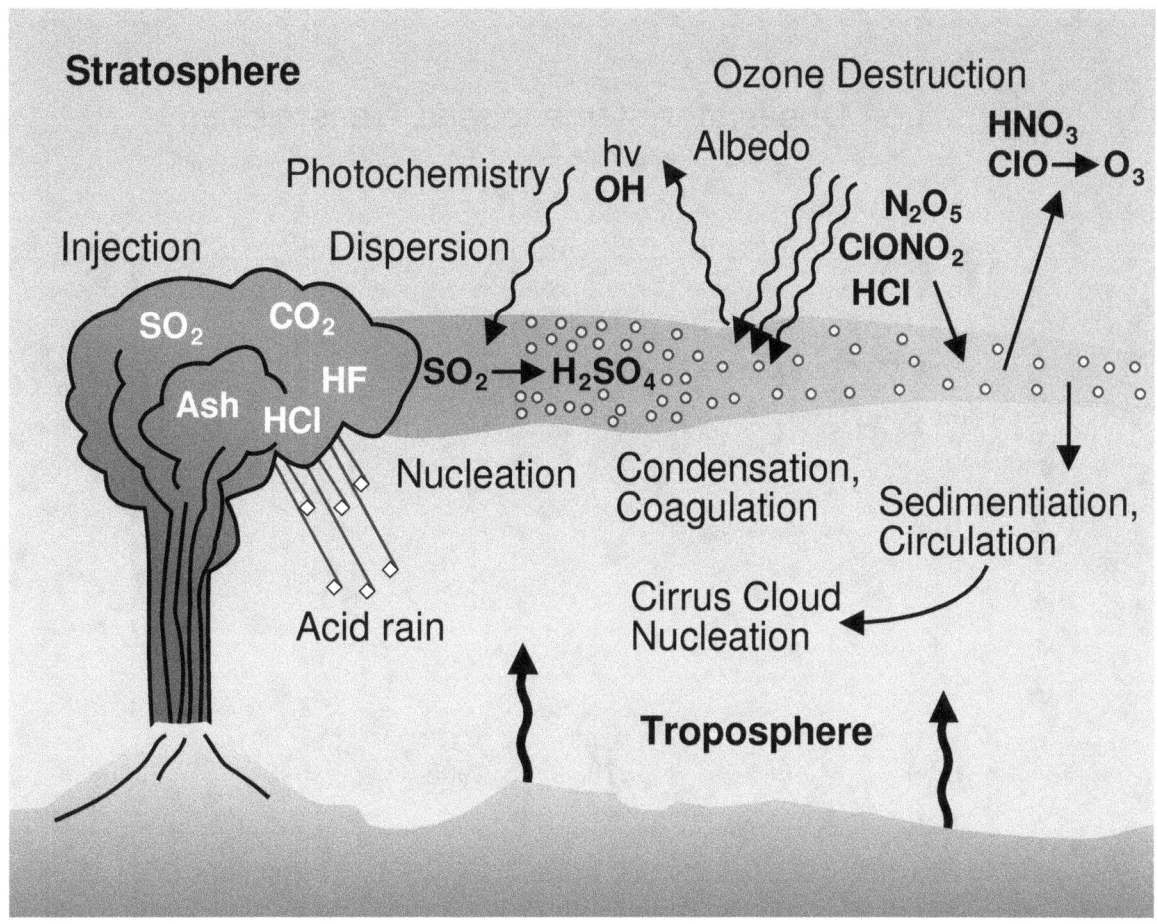

Schematic of volcano injection of aerosols and gases

Mauna Loa Observatory Atmospheric Transmission

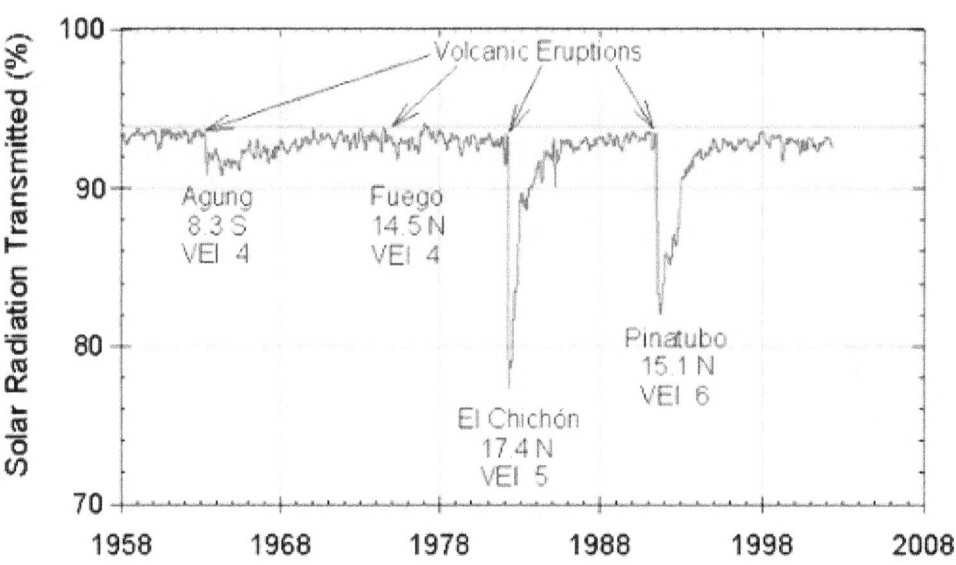

Solar radiation graph 1958–2008, showing how the radiation is reduced after major volcanic eruptions

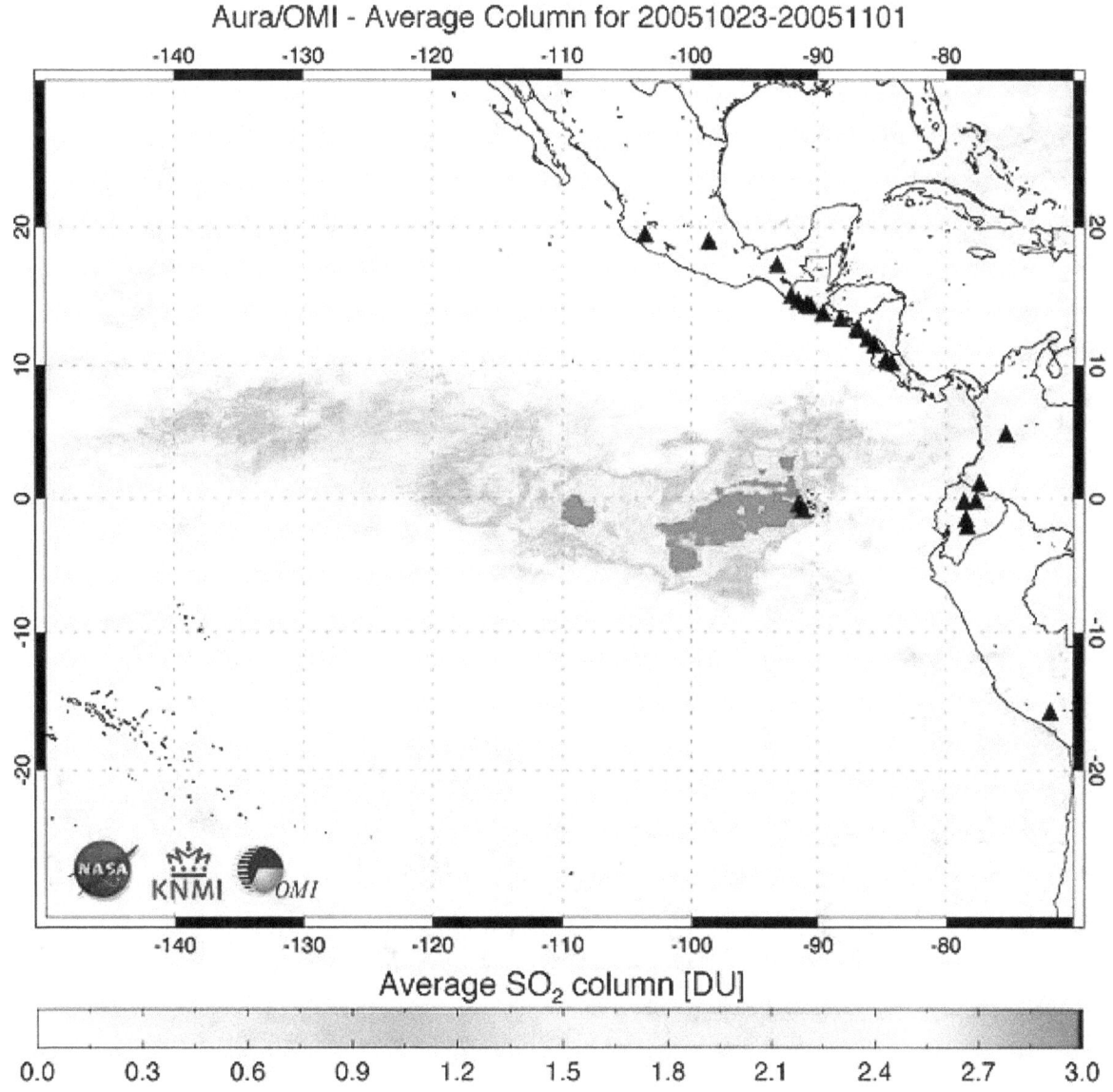

Sulfur dioxide concentration over the Sierra Negra Volcano, Galapagos Islands, during an eruption in October 2005

Ash plume rising from Eyjafjallajökull on April 17, 2010

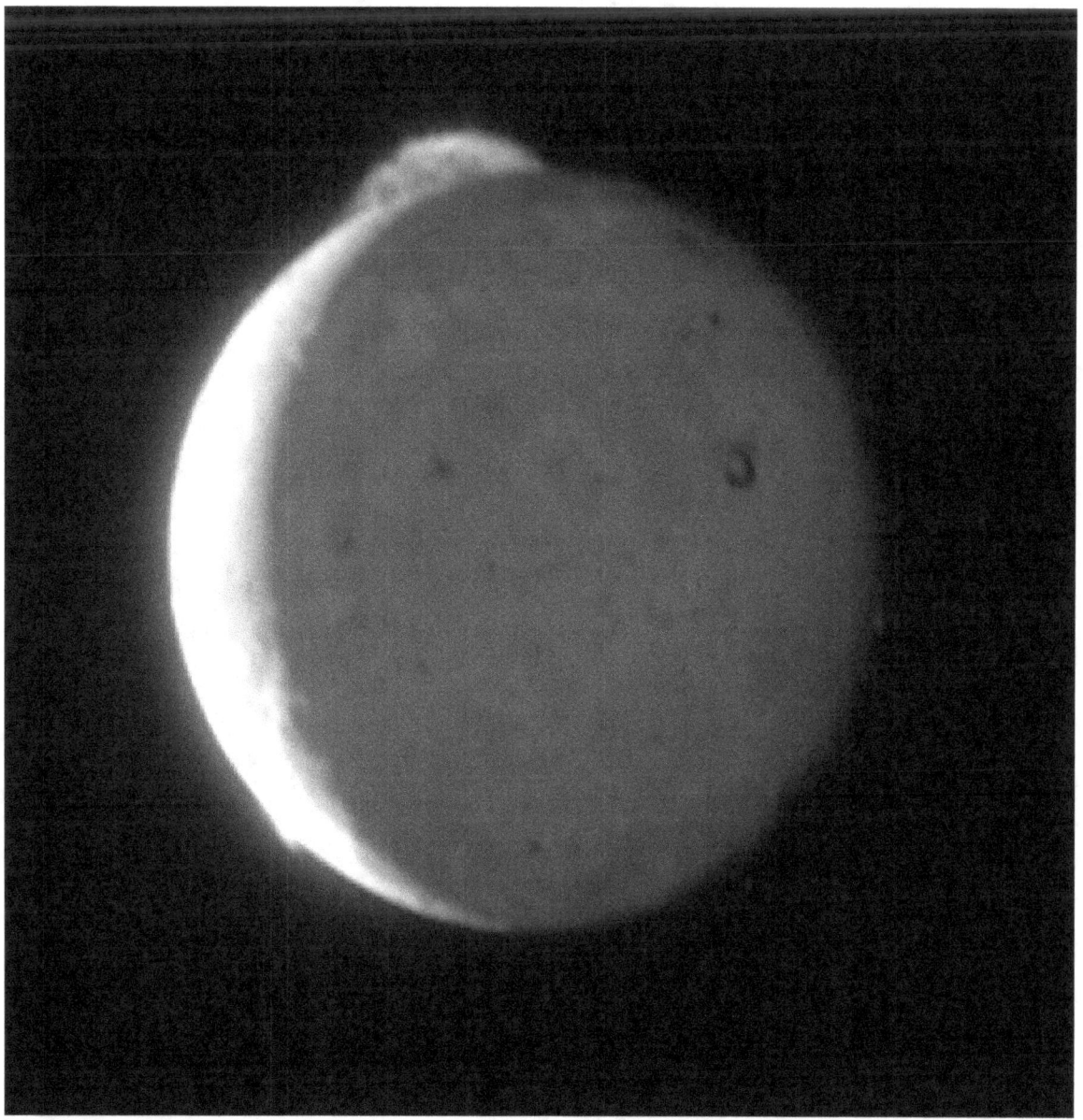

The Tvashtar volcano erupts a plume 330 km (205 mi) above the surface of Jupiter's moon Io.

Chapter 2

Seamounts of the Atlantic Ocean

2.1 Allegheny Seamount

Coordinates: 36°52′7.6″N 58°44′16.4″W / 36.868778°N 58.737889°W The **Allegheny Seamount** is a seamount in the Atlantic Ocean. It is part of the New England Seamount chain, which was active more than 100 million years ago. It was formed when the North American Plate moved over the New England hotspot.[*][1]

2.1.1 References

[1] "Geological Origin of the New England Seamount Chain". *National Oceanic and Atmospheric Administration*. U.S. Department of Commerce. 2005. Retrieved 2007-09-16.

2.2 Anton Dohrn Seamount

The **Anton Dohrn Seamount** is a guyot in the Rockall Trough in the northeast Atlantic. It was named after the German fishery research vessel which discovered it at the end of the 1950s which, in turn, had been named after the 19th-century biologist Anton Dohrn.

The feature rises from approximately 2,100 metres to 600 metres below sea level and has a sedimentary layer approximately 100 metres thick. It arose through episodic volcanic activity between 70 and 40 million years ago.[*][1]

Around the base of the seamount is a slight "moat" where the sea-bottom is at a lower depth than the surrounding terrain.

2.2.1 References

[1] O'Connor, Stofferes, Wijbrans, Shannon and Morrissey (2000). Evidence from episodic seamount volcanism for pulsing of the Iceland plume in the past 70 Myr, *Nature* 408, 954-958.

2.2.2 External links

- WWF report on seamounts of the Northeast Atlantic (PDF)

2.3 Asterias Seamount

Coordinates: 38°53′56″N 65°17′59.8″W / 38.89889°N 65.299944°W The **Asterias Seamount** is a seamount in the

Atlantic Ocean. It is part of the New England Seamount chain, which was active more than 100 million years ago. It was formed when the North American Plate moved over the New England hotspot.*[1]

2.3.1 References

[1] "Geological Origin of the New England Seamount Chain". *National Oceanic and Atmospheric Administration*. U.S. Department of Commerce. 2005. Retrieved 2007-09-16.

2.4 Balanus Seamount

Coordinates: 39°22′58.8″N 65°22′47.3″W / 39.383000°N 65.379806°W

The **Balanus Seamount** is a seamount in the Atlantic Ocean. It is part of the New England Seamount chain, which was active more than 100 million years ago. It was formed when the North American Plate moved over the New England hotspot.*[1]

2.4.1 References

[1] "Geological Origin of the New England Seamount Chain". *National Oceanic and Atmospheric Administration*. U.S. Department of Commerce. 2005. Retrieved 2007-09-16.

2.5 Bean Seamount

The **Bean Seamount** is a small seamount in the northern Atlantic Ocean. It is part of the Corner Rise Seamounts, which was active about 75 million years ago. It was formed when the North American Plate moved over the New England hotspot.*[2]

2.5.1 References

[1] http://www.whoi.edu/oceanus/viewImage.do?id=57092&aid=33769

[2] "A Hundred-Million Year History of the Corner Rise and New England Seamounts". *National Oceanic and Atmospheric Administration*. U.S. Department of Commerce. 2005. Retrieved 2009-04-11.

2.6 Bear Seamount

The **Bear Seamount** is a guyot or flat-topped underwater volcano in the Atlantic Ocean. It is the oldest of the New England Seamount chain, which was active more than 100 million years ago. It was formed when the North American Plate moved over the New England hotspot.*[3]

2.6.1 Formation

The Bear Seamount is the first guyot in a chain of about 30 extinct volcanoes extending in a straight line south-eastwards from the edge of the continental shelf near Woods Hole, Massachusetts to north-east of Bermuda. These seamounts resulted from the movement of a mantle plume hotspot. This hotspot is now under the Great Meteor Seamount. The chain rises about 4,000 metres (13,000 ft) above the surrounding Sohm Abyssal Plain. Over time they have been eroded and have developed flat table-like summits surrounded by slopes with an inclination of about 20°. The currents in the vicinity of the Bear Seamount include the warm water Gulf Stream flowing towards the north east, the deep boundary current flowing along the continental shelf towards the south west, and the deep, icy cold Antarctic bottom water that flows past the lower flanks of the chain.*[4]

Bear Seamount rises approximately 2,000 to 3,000 metres (6,600 to 9,800 ft) above the surrounding seabed and the roughly flat summit is about 1,100 metres (3,600 ft) below the surface of the sea. The top is covered by a deep layer of sediment through which basaltic rocks and erratic boulders protrude. Much of this material has fallen from above, probably from icebergs that drifted southwards during the Pleistocene.*[4]

2.6.2   Biodiversity

Because little was known about the biodiversity of the New England Seamount Chain, an expedition was mounted in 2000. The NOAA National Marine Fisheries Service deep water research vessel R/V Delaware II made 20 exploratory trawls in the vicinity of Bear Seamount and around 274 species were collected. These included 115 species of fish, some of which were rare or had not been recorded in the western North Atlantic before. The roundnose grenadier (*Coryphaenoides rupestris*) and the onion-eye grenadier (*Macrourus berglax*) were the only fish species of potential commercial importance – they were caught in mid-water at depths of between 1,100 and 1,800 metres (3,600 and 5,900 ft) and were up to a metre in length.*[4] A common but much smaller fish was *Aldrovandia phalacra*.*[4]

Twenty-six species of cephalopods were collected including squid such as *Mastigoteuthis agassizii* and *Mastigoteuthis magna*. Other invertebrates caught by trawls dragged along the seamount surface included 46 species of crustacean such as the prawns *Sergestes spp.* and *Acanthephyra spp.*, and the shrimp *Pasiphaea spp.* Also present was the bank-forming deepwater coral *Lophelia pertusa*, which supported a community of worms, hydroids and other corals. Brittle stars, especially *Ophiomusium lymani*, were numerous as were the sea urchin *Echinus affinis*, the sea star *Neomorphaster forcipatus*, mysids and various scyphozoans.*[4]

2.6.3   Notes and references

[1] "NOAA Ocean Explorer". Retrieved 2007-08-13.

[2] "Alvin Dive Information". Retrieved 2007-08-13.

[3] "Geological Origin of the New England Seamount Chain". *National Oceanic and Atmospheric Administration*. U.S. Department of Commerce. 2005. Retrieved 2007-09-16.

[4] Moore, J.A.; Vecchione, M.; Collette, B. B.; Gibbons, R.; Hartel, K. E.; Galbraithe, J. K.; Turnipseed, M.; Southworth, M.; Watkins, E. (2003). "Biodiversity of Bear Seamount, New England Seamount Chain: Results of exploratory trawling" (PDF). *Journal of Northwest Atlantic Fishery Science* **31**: 363–372.

2.6.4   External links

- NOAA: Animated fly-through of the Bear Seamount

2.7   Buell Seamount

The **Buell Seamount** is a seamount in the Atlantic Ocean. It is part of the New England Seamount chain, which was active more than 100 million years ago. It was formed when the North American Plate moved over the New England hotspot.*[1]

2.7.1   References

[1] "Geological Origin of the New England Seamount Chain". *National Oceanic and Atmospheric Administration*. U.S. Department of Commerce. 2005. Retrieved 2007-09-16.

Coordinates: 39°3′46.6″N 66°24′0.3″W / 39.062944°N 66.400083°W

2.8 Caloosahatchee Seamount

The **Caloosahatchee Seamount** is a seamount in the northern Atlantic Ocean. It is part of the Corner Rise Seamounts, which was active about 75 million years ago. It was formed when the North American Plate moved over the New England hotspot.*[2]

2.8.1 References

[1] http://earthref.org/SC/SMNT-347N-0498W/ Caloosahatchee Seamount at Earthref.org

[2] "A Hundred-Million Year History of the Corner Rise and New England Seamounts". *National Oceanic and Atmospheric Administration*. U.S. Department of Commerce. 2005. Retrieved 2009-04-11.

2.9 Caryn Seamount

The **Caryn Seamount** is a seamount in the Atlantic Ocean. It is an independent seamount located southwest of the New England Seamount chain, which was active more than 100 million years ago.*[1]

2.9.1 References

[1] http://earthref.org/cgi-bin/sc.cgi?id=SMNT-367N-0680W

Coordinates: 36°40.8′N 67°57′W / 36.6800°N 67.950°W

2.10 Corner Rise Seamounts

The **Corner Rise Seamounts** are a chain of extinct submarine volcanoes in the northern Atlantic Ocean east of the New England Seamount chain. Both it and the New England Seamount Chain were formed by the Great Meteor hotspot.*[1] It is the shallowest seamount in New England, with some of its nineteen highest peaks only 800–900 m deep.*[2]

Like most seamounts, they attracts fish. Over 175 species have been found there,*[1] including splendid alfonsino, black cardinal fish, black scabbardfish, and wreckfish.*[2] Trawl fishing during the 1970s and 1980s resulted in approximately 20,000 tons of fish being harvested.*[1] As a result, the seamounts were closed to demersal fishing (collecting fish near the bottom of the ocean, as opposed to pelagic fishing, collecting fish near the surface) beginning 1 January 1997. The original ban was supposed to be lifted 31 December 2010,*[1] but was extended until 31 December 2020.*[2] Almost a decade into the ban, a 2005 Woods Hole Oceanographic Institution survey found that two of the peaks, Kükenthal and Yakutat, had been stripped bare of both corals and bottom-dwelling animals.*[3]*[4] However the survey, which covered both the Corner Rise and New England seamounts, found 270 species of invertebrates and crustaceans, including 70 species unique to the Corner Rise Seamounts.*[5]

2.10.1 Seamounts

Seamounts within the Corner Rise Seamount chain include:

- Bean Seamount
- Caloosahatchee Seamount with
  - Milne-Edwards Peak
  - Verrill Peak

- Castle Rock Seamount

- Corner Seamount with

    - Goode Peak
    - Kukenthal Peak

- Justus Seamount

- MacGregor Seamount

- Rockaway Seamount

- Yakutat Seamount

2.10.2 References

[1] Shank, Timothy M. (March 2010). "SPOTLIGHT 4: New England and Corner Rise Seamounts" (PDF). *Oceanography* **23** (1). Retrieved 18 August 2015.

[2] "Corner Seamounts". *Food and Agriculture Organization of the United Nations*. Northwest Atlantic Fisheries Organization. Retrieved 18 August 2015.

[3] Kusek, Kristen M. (6 November 2007). "Coral Catastrophe on the Corner Rise Seamounts". *Oceanus Magazine* **46** (2). Woods Hole Oceanographic Institute. Retrieved 18 August 2015.

[4] Waller, Rhian & Les Watling (1 October 2007). "Anthropogenic Impacts on the Corner Rise Seamounts, North-West Atlantic Ocean". *DigitalCommons @ UMaine*. University of Maine. Retrieved 19 August 2015.

[5] Auster, Peter J. "Linking Biodiversity in the Deep Sea to International Management Needs" (PDF). *Oceans and Law of the Sea*. United Nations. Retrieved 18 August 2015.

2.10.3 External links

- "Seamount Catalog". Seamount Biogeosciences Network. Retrieved 18 August 2015.

- "Image of New England and Corner Rise Seamounts". Gulf of Maine Census. Retrieved 19 August 2015.

2.11 Dom João de Castro Bank

Dom João de Castro Bank (Portuguese: *Banco de D. João de Castro*) is a large submarine volcano located in the central north Atlantic Ocean, between the islands of São Miguel and Terceira in the archipelago of the Azores.

2.11.1 History

The first historical reference correlated with the perilous submarine volcanoes between São Miguel and Terceira occurred from the sinking of two ships of the French corsair Henry Tourin, in the spring of 1718.[1]

The last major eruption associated with this region occurred on 31 December 1720.[2] Beginning as a submarine eruption, it eventually built to a Surtseyan eruption that resulted in the formation of a circular island 1.5 kilometres (0.93 mi) long, 250 m (800 ft) altitude island.[2]

Designated the *Ilha Nova*, it remained above sea-level for only two years, reaching a height of 180 metres (590 ft) and diameter of 900 metres (3,000 ft). Marine erosion and ocean swells reduced and degraded the cone's size considerably, so that by 21 July 1722 the *Conselho da Marinha Português* was advised that island had disappeared. (Oliveira, 1943).[1][2]

After two centuries of disputed references to the existence of the island, on 28 July 1941 the hydrographic ship *NH D. João de Castro* identified its position and cataloged the morphology of the bank. In keeping with nautical tradition at its discovery, the bank inherited its name.*[1]*[2]

The most recent seismic events from the region occurred from 4 June 2012, when incrementally increasing tremors affected the region.*[3] A maximum seismic event was recorded from the islands of São Miguel and Terceira, equivalent to levels III/IV on the Mercalli Scale.*[3]

2.11.2 Geography

It remains a seismically active zone, lying midway between the islands of São Miguel and Terceira, rising to within 14 m (46 ft) of the sea surface.*[2] The volcano has a large fumarole field, its top approximately 12 metres (39 ft) from the surface and 1,600 metres (5,200 ft) in total.

Two parasitic craters, both about 90 by 45 metres (295 ft × 148 ft) wide, are located on the northwest flank. The younger crater displays a floor consisting of a chilled lava lake with polygonal surface fractures, while the older, less distinct crater is obscured by tephra deposits.

Biome

D. João de Castro is an important fishing ground for both demersal fish, such as the black seabream (*Spondyliosoma cantharus*) and blackbelly rosefish, (*Helicolenus dactylopterus*), as well as pelagic tuna species (**Thunnus** and others).

On 27 July 1996, in a scientific expedition organized by the *Clube Naval de Ponta Delgada*, marine biology investigatores from the University of the Azores completed a study to identify species localized within the bank.*[2]

2.11.3 See also

- List of volcanoes in Azores

2.11.4 References

Notes

[1] Nunes, João Carlos (23 June 2013), Lusa/AO Online, ed., *Banco D. João de Castro*, Ponta Delgada (Azores), Portugal: Açoriano Oriental, p. 20

[2] Sérgio Paulo Ávila (1997), p.1

[3] Lusa/AO Online, ed. (13 June 2012), *Banco D. João de Castro voltou a registar atividade sísmica* (in Portuguese), Ponta Delgada (Azores), Portugal: Açoriano Oriental

Sources

- "Don Joao de Castro Bank". *Global Volcanism Program*. Smithsonian Institution.

- Scarth, Alwyn; Tanguy, Jean-Claude (2001). *Volcanoes of Europe*. Oxford University Press. p. 243 pp. ISBN 0-19-521754-3.

- Ávila, Sérgio Paulo, *Moluscos Marinhos Recolhidos no Banco "D. João de Castro"* (PDF) (in Portuguese) (3) (8 ed.), Horta (Azores), Porutgal: Açoreana, pp. 331–332

- Oliveira, A. (1943), "Banco "D. João de Castro" ", *Trabalhos da Missão Hidrográfica das Ilhas Adjacentes* (in Portuguese), Lisbon, Portugal, pp. 1–19

2.12   Ewing Seamount

Coordinates: 23°14′40″S 8°16′20″E / 23.24444°S 8.27222°E The **Ewing Seamount** is a seamount in the southern Atlantic Ocean, which lies on the Tropic of Capricorn.[*][1][*][2][*][3] Ewing is part of the Walvis Ridge[*][1] having a mean depth of 4,500 metres and a summit depth of 700 metres.[*][4]

2.12.1   References

[1] National Geographic Atlas of the World: Revised Sixth Edition, National Geographic Society, 1992

[2] encarta.msn.com

[3] earthref.org

[4] Seamounts online

2.13   Fogo Seamounts

The **Fogo Seamounts**, also called the **Fogo Seamount chain**, are a group of seamounts located about 500 km (311 mi) offshore of Newfoundland and southwest of the Grand Banks of Newfoundland. They consist of basaltic submarine volcanoes that formed during the Early Cretaceous period.[*][1]

The volcanic activity that formed the Fogo Seamounts could have originated in two ways. They may have formed as a result of magma rising along a linear fault zone or from the North American Plate passing over the Canary or Azores hotspots.[*][1]

2.13.1   See also

- Volcanism of Canada
- Volcanism of Eastern Canada
- List of volcanoes in Canada

2.13.2   References

[1] Morphology, petrography, age and origin of Fogo Seamount chain, offshore eastern Canada

Coordinates: 41°45′47.4″N 52°16′51.6″W / 41.763167°N 52.281000°W

2.14   George Bligh Bank

Coordinates: 58°55′N 14°45′W / 58.917°N 14.750°W

George Bligh Bank is a seamount in the northeast Atlantic, west of Scotland. It lies in Rockall Trough, to the north of Rockall Bank.

2.15   Gerda Seamount

The **Gerda Seamount** is a seamount in the Atlantic Ocean. It is part of the New England Seamount chain, which was active more than 100 million years ago. It was formed when the North American Plate moved over the New England hotspot.[*][1]

2.15.1 References

[1] "Geological Origin of the New England Seamount Chain". *National Oceanic and Atmospheric Administration*. U.S. Department of Commerce. 2005. Retrieved 2007-09-16.

2.16 Gilliss Seamount

The **Gilliss Seamount** is a seamount in the Atlantic Ocean. It is part of the New England Seamount chain, which was active more than 100 million years ago. It was formed when the North American Plate moved over the New England hotspot.[*][1]

2.16.1 References

[1] "Geological Origin of the New England Seamount Chain". *National Oceanic and Atmospheric Administration*. U.S. Department of Commerce. 2005. Retrieved 2007-09-16.

2.17 Gorringe Ridge

The **Gorringe Ridge** is a seamount in the Atlantic Ocean. It is located about 130 miles (210 km) west of Portugal, between the Azores and the Strait of Gibraltar along the Azores–Gibraltar fault zone. It is about 60 km wide and 180 km long in the northeast direction.

2.17.1 Discovery

In the nineteenth century the United States Coast Survey embarked on an ambitious program to map the seafloors of the world's main oceanways. This produced extensive maps of the more shallow areas, but deep-ocean work was hampered by lack of robust equipment. In 1872, English scientist Sir William Thomson invented a wire-based depth-sounding mechanism which was a significant improvement over rope-type equipment used previously. This Thomson Sounding Machine made its first discovery in 1874, of several seamounts west of the Hawaiian Islands.[*][2] Its second use was on the USS *Gettysburg* (1858), an ocean-going vessel used in 1875 to extensively map the Eastern Atlantic seafloor. The ship was commanded by Captain Henry Honychurch Gorringe. On 6 November 1875 this expedition discovered the raised area (which was referred to as *Gorringe Bank* in reference to the ship's captain), and spent time mapping it. They determined that it contained two significant peaks, which they named **Gettysburg** (the highest, at 20 meters depth) and **Ormonde** (the second highest, at 33 meters depth).[*][3]

2.17.2 Subsequent explorations

In the early twentieth century, Albert I, Prince of Monaco spent considerable time exploring and mapping the Gorringe Bank, using a total of three ships: *Princess Alice*, *Princess Alice II*, and *Hirondelle II*. The ships' names were given to several mounds and large banks between Madeira and the Azores.[*][4]

In June 2005, the Oceana Organization mounted an extensive exploration of the biota on Gorringe Ridge's two largest peaks. It aims to categorize and determine relative abundance of the diverse lifeforms there.[*][5]

2.17.3 Geology

The Gorringe Bank was eventually renamed *Gorringe Ridge* owing to its extensive length and the determination that it is the result of two tectonic plates which are sliding into and past each other. The plate boundaries here are converging at 4 mm/y, as well as sliding past each other. Upper mantle and oceanic crust are exposed along this ridge. Ferrogabbro dated at 77 Mya has been intruded, Also at 66 Mya the Canary hotspot mantle plume passed by and caused alkaline magma to intrude. Where there is crust, it is very thin, so that the Moho comes up to the sea floor. Sediment overlies the mantle, so this could be considered as crust. Since the Miocene Era there has been shortening of the ocean crust absorbed by folding, and thrusting.

A 2003 study[6] of the ridge's gravity and magnegic anomalies concluded that the Moho is relatively flat across the ridge, and that the ridge's upper part corresponds to a northwestwards vergent fold. The thrusting activity probably started some 20 million years ago, and has covered about 20 km. The seamount is composed of gabbros of the oceanic crust, serpentized rocks and alkaline basalts.[7]

2.17.4   1755 Lisbon earthquake

Modern seismologists who studied the cause of the 1755 Lisbon earthquake and the resulting tsunami initially suspected a displacement in the Gorringe Ridge, but later concluded that there was a simultaneous event involving two separate faults along the African Plate boundary, both faults displacing by around 20m.[8]

2.17.5   See also

- Geology of the Iberian Peninsula

2.17.6   References

[1] "Seamounts Catalog" . Earthref, a National Science Foundation project. Retrieved 28 December 2009.

[2] http://oceanexplorer.noaa.gov/history/timeline/timeline.html NOAA Timeline (website), accessed 16 August 2009

[3] Continental Lithospheric Contribution to Alkaline Magmatism: Isotopic (Nd, Sr, Pb) and Geochemical (REE) Evidence from Serra de Monchique and Mount Ormonde Complexes

[4] http://oceana.org/europe/publications/reports/seamounts-of-the-brgorringe-bank/#4352 Oceana website, accessed 16 August 2009

[5] *Oceana website*

[6] Geophysical International Journal, ISSN 0956-540X (2003) **153**, 3, pp. 586–594

[7] http://cat.inistfr/?aModele=afficheN&cpsidf=14837494

[8] "The Great Earthquake 1755" . bbc.co.uk. 28 April 2004. Retrieved 28 April 2009.

2.17.7   External links

- "We depart towards the Gorringe Ridge" . Oceania.org.

2.18   Gosnold Seamount

The **Gosnold Seamount** is a seamount in the Atlantic Ocean. It is part of the New England Seamount chain, which was active more than 100 million years ago. It was formed when the North American Plate moved over the New England hotspot.[1]

2.18.1   References

[1] "Geological Origin of the New England Seamount Chain". *National Oceanic and Atmospheric Administration*. U.S. Department of Commerce. 2005. Retrieved 2007-09-16.

2.19 Great Meteor Seamount

The **Great Meteor Seamount** is a large guyot (tablemount) located south of the Azores in the Atlantic Ocean. It is the New England hotspot's most recent eruptive center and is one of the most completely investigated seamounts in the world. This guyot rises up from a depth of almost 4,800 meters to about 270 meters below the surface of the Atlantic. The summit measures 50 km x 28 km (1465 km^2).[1] On the southwest side it is flanked by the smaller Little Meteor Seamount[1] and the still smaller Closs Seamount.[2]

The German research vessel *Meteor* discovered the tablemount between 1925 and 1927. It was given the name Great Meteor *Bank*, a designation still used in the official GEBCO gazetteer.

2.19.1 Formation

The New England hotspot formed the White Mountains 124 to 100 million years ago when the North American continent was directly overhead. As the continent drifted to the west, the hotspot gradually moved offshore. On a southeasterly course, the hotspot formed Bear Seamount, the oldest in the chain, about 100 to 103 million years ago. Over the course of millions of years, it continued creating the rest of the seamounts, eventually culminating in the Nashville Seamount about 83 million years ago. As the Atlantic Ocean continued to spread, the hotspot eventually "traveled" further east, forming the Great Meteor Seamount where it is found today.[3]

2.19.2 Ecology

The unique ecological condition of the Great Meteor Seamount is shown by the many endemic copepod and nematode species.[4]

2.19.3 References

[1] METEOR Retrieved on 2007-10-04

[2] Verhoef, Jaap (1984). "A Geophysical Study of the Atlantis-Meteor Seamount Complex". *Geologica Ultraiectina* **38**: 1 – 153. ISSN 0072-1026. Retrieved 14 April 2014.

[3] "Geological Origin of the New England Seamount Chain". *Office of Ocean Exploration and Research, National Oceanic and Atmospheric Administration*. U.S. Department of Commerce. 2005. Retrieved 14 April 2014.Revised August 25, 2010 by the Ocean Explorer

[4] SPOTLIGHT 5 *Great Meteor Seamount*

2.20 Gregg Seamount

The **Gregg Seamount** is a seamount in the Atlantic Ocean. It is part of the New England Seamount chain, which was active more than 100 million years ago. It was formed when the North American Plate moved over the New England hotspot.[1]

2.20.1 References

[1] "Geological Origin of the New England Seamount Chain". *National Oceanic and Atmospheric Administration*. U.S. Department of Commerce. 2005. Retrieved 2007-09-16.

2.21   Hodgson Seamount

The **Hodgson Seamount** is a seamount in the Atlantic Ocean. It is part of the New England Seamount chain which was active more than 100 million years ago. It was formed when the North American Plate moved over the New England hotspot.*[1] It was named after Robert David Hodgson, an American geographer and an internationally recognized expert on geographic aspects of the law of the sea and maritime boundaries.

2.21.1   References

[1] "Geological Origin of the New England Seamount Chain". *National Oceanic and Atmospheric Administration*. U.S. Department of Commerce. 2005. Retrieved 2007-09-16.

2.22   Kelvin Seamount

The **Kelvin Seamount** is a guyot in the Atlantic Ocean. It is part of the New England Seamount chain, which was active more than 100 million years ago. It was formed when the North American Plate moved over the New England hotspot.*[1]

2.22.1   Notes and references

[1] "Geological Origin of the New England Seamount Chain". *National Oceanic and Atmospheric Administration*. U.S. Department of Commerce. 2005. Retrieved 2007-09-16.

2.23   Kiwi Seamount, Atlantic Ocean

The **Kiwi Seamount** is a seamount in the Atlantic Ocean. It is part of the New England Seamount chain, which was active more than 100 million years ago. It was formed when the North American Plate moved over the New England hotspot.*[1]

2.23.1   References

[1] "Geological Origin of the New England Seamount Chain". *National Oceanic and Atmospheric Administration*. U.S. Department of Commerce. 2005. Retrieved 2007-09-16.

2.24   Manning Seamount

The **Manning Seamount** is a guyot in the Atlantic Ocean. It is part of the New England Seamount chain, which was active more than 100 million years ago. It was formed when the North American Plate moved over the New England hotspot.*[1]

2.24.1   Notes and references

[1] "Geological Origin of the New England Seamount Chain". *National Oceanic and Atmospheric Administration*. U.S. Department of Commerce. 2005. Retrieved 2007-09-16.

2.25 Michael Seamount

The **Michael Seamount** is a seamount in the Atlantic Ocean. It is part of the New England Seamount chain, which was active more than 100 million years ago. It was formed when the North American Plate moved over the New England hotspot.*[1]

2.25.1 References

[1] "Geological Origin of the New England Seamount Chain". *National Oceanic and Atmospheric Administration*. U.S. Department of Commerce. 2005. Retrieved 2007-09-16.

2.26 Muir Seamount

Muir Seamount is a seamount (underwater volcano), located at 33°43.20′N 62°29.40′W / 33.72000°N 62.49000°WCoordinates: 33°43.20′N 62°29.40′W / 33.72000°N 62.49000°W.*[1] It is located in the Bermuda rise, a seismically active region, and was the site of an earthquake on March 24, 1978.*[2]

Fourteen core samples of up to 125,000 years of age were collected off of Muir Seamount, ranging in depth to more than 3,000 m (9,843 ft). These samples were used to analyze the weight percentage of carbonate over time on the ocean floor, which in turn would provide clues to the oceanic environment over time. Overall concentrations were most stable at 4,650 m (15,256 ft) of depth. This was interpreted as indicating an increase in flow form colder waters of Antarctic flow, and the stagnation of the North American flows.*[3] A total of 8 core samples have been gathered, with ages ranging from Pliocene to Late Cretaceous. No radiometric dates are available for the seamount, but it may of Early Cretaceous time.*[4]

Over 150 invertebrates have been collected in total from Bear, Manning, and Muir seamounts. Ophiuroids and goose barnacles are the most abundant animal found on the seamount, as is *Asteroshchema sp.*, a basket star. Paragorgia have also been observed. Genetic tests on the specimens collected indicated that Muir Seamount is the most genetically isolated of the four.*[5]

2.26.1 References

[1] "Seamount Catalog: Muir Seamount". *Seamounts database*. EarthRef, a National Science Foundation project. Retrieved 2013-01-16.

[2] Nishenko, S. P.; Kafka, A.L. (1982). "Earthquake focal mechanisms and the intraplate setting of the Bermuda rise". *Journal of Geophysical Research* **87** (B5): 3929–3942. Bibcode:1982JGR....87.3929N. doi:10.1029/JB087iB05p03929. ISSN 0148-0227.

[3] L. Balsam, William (September 1983). "Carbonate dissolution on the Muir Seamount (western North Atlantic); interglacial/glacial changes". *Journal of Sedimentary Research* (SEPM [Society for Sedimentary Geology]) **53** (3): 719–731. doi:10.1306/212F82AB-2B24-11D7-8648000102C1865D. Retrieved 10-12-09. Check date values in: |accessdate= (help)

[4] Gillain R. Foulger and Donna M. Jurdy, ed. (2007). *Plates, plumes, and planetary processes*. Special Paper 430. Geological Society of America. p. 562. ISBN 978-0-8137-2430-0.

[5] Walter Cho and Dr. Timothy M. Shank. "Molecular Systematics and Population Connectivity of Seamount Faunal Populations on the New England and Corner Rise Seamounts". Woods Hole Oceanographic Institution (WHOI). Retrieved 2013-01-16.

2.27 Mytilus Seamount

The **Mytilus Seamount** is a seamount in the Atlantic Ocean. It is part of the New England Seamount chain, which was active more than 100 million years ago. It was formed when the North American Plate moved over the New England hotspot.*[1]

2.27.1   References

[1] "Geological Origin of the New England Seamount Chain". *National Oceanic and Atmospheric Administration*. U.S. Department of Commerce. 2005. Retrieved 2007-09-16.

2.28   Nashville Seamount

The **Nashville Seamount** is a seamount in the Atlantic Ocean. It is part of the New England Seamount chain, which was active more than 100 million years ago. It was formed when the North American Plate moved over the New England hotspot.*[1]

2.28.1   References

[1] "Geological Origin of the New England Seamount Chain". *National Oceanic and Atmospheric Administration*. U.S. Department of Commerce. 2005. Retrieved 2007-09-16.

2.29   New England Seamount chain

The **New England Seamount chain** is an underwater chain of seamounts in the Atlantic Ocean stretching over 1,000 km from the edge of the Georges Bank off the coast of Massachusetts. The chain consists of over twenty extinct volcanic peaks, many rising over 4,000 m from the seabed.*[1]*[2] It is the longest seamount chain in the North Atlantic and harbours a diverse range of deep sea fauna.*[2] Scientists have visited the chain on various occasions to survey the geologic makeup and biota of the region. The chain forms part of the Great Meteor hotspot track, having formed by the movement of the North American Plate over the New England hotspot. The oldest volcanoes that were formed by the same hotspot are northwest of Hudson Bay, Canada.

2.29.1   Formation

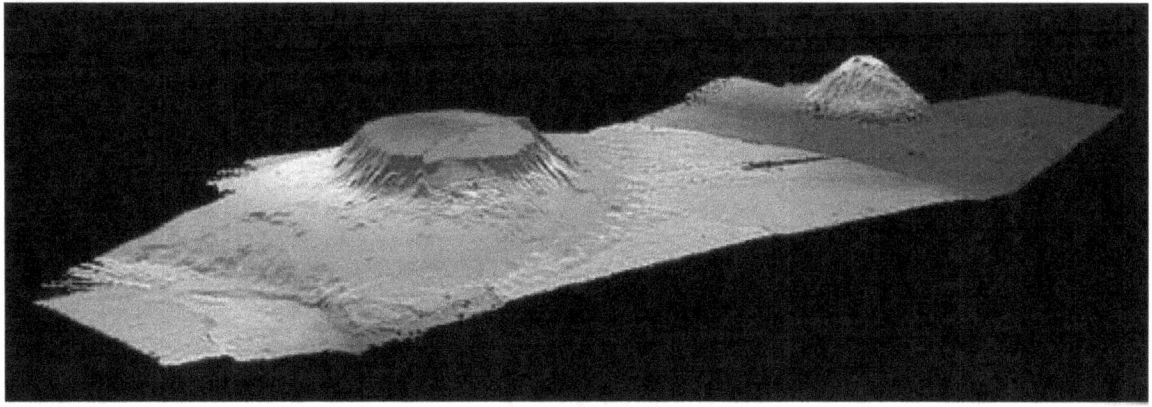

Bear Seamount

The New England hotspot, also referred to as the Great Meteor hotspot, formed the White Mountains 124 to 100 million years ago when the North American continent was directly overhead. As the continent drifted to the west, the hotspot gradually moved offshore. On a southeasterly course, the hotspot formed Bear Seamount, the oldest in the chain, about 100 to 103 million years ago. Over the course of millions of years, it continued creating the rest of the seamounts, eventually culminating in the Nashville Seamount about 83 million years ago. As the Atlantic Ocean continued to spread, the hotspot eventually "travelled" further east, forming the Great Meteor Seamount south of the Azores, where it is found today.[*][3] The New England Seamounts were once at or above sea level. As time passed, however, and the chain moved farther away from the New England hotspot, the crust cooled and contracted, sinking back down to the ocean. The peaks are now all a km or more below the surface.

Some animals from the New England Seamounts: gorgonian soft coral, a brisingid sea star, and sponges

2.29.2 Biota

The seamount chain provides a unique habitat for deep sea marine creatures. Coral formations grow on the rocky outcrops, resembling underwater forests that provide shelter for invertebrates and fish.[*][4] Due to the expenses and difficulties of studying the deep ocean, little was known of the creatures that inhabited the New England Seamounts. In fact, before recent expeditions, there was only one known coral species in the entire chain.[*][2] Marine biologists caught and classified over 203 species of fish and 214 species of invertebrates on the Bear Seamount in various exploratory studies since 2000.[*][2] This range of diversity suggests that other seamounts may harbour more unknown macro-organisms. In fact, during one survey, a species of cutthroat eel believed to be found only near Australia was identified.[*][5] Corals, echinoderms, and crustaceans make up a large portion of the creatures found on the seamount. These organisms act as indicator species, identifying potential problems in the ecosystem.[*][2]

2.29.3 Seamounts

Seamounts within the New England Seamount chain include:

- Allegheny Seamount

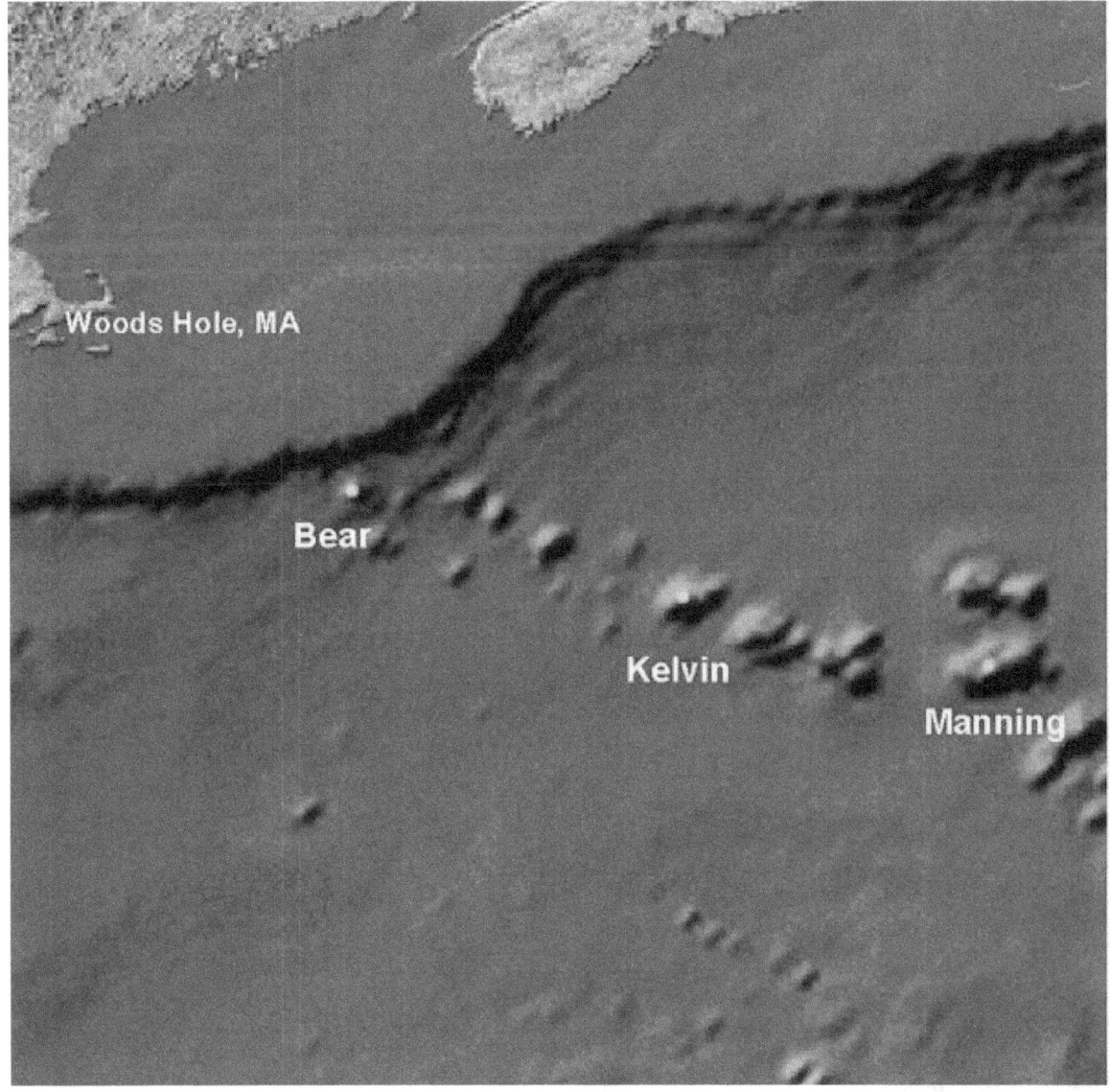

Map of the New England Seamount chain showing the locations of Bear, Kelvin and Manning seamounts

- Asterias Seamount

- Balanus Seamount

- Bear Seamount

- Buell Seamount

- Gerda Seamount

- Gilliss Seamount

- Gosnold Seamount

- Gregg Seamount

- Hodgson Seamount

- Kelvin Seamount

- Kiwi Seamount

- Manning Seamount

- Michael Seamount

- Mytilus Seamount

- Nashville Seamount

- Panulirus Seamount

- Picket Seamount

- Physalia Seamount

- Rehoboth Seamount

- Retriever Seamount

- San Pablo Seamount

- Sheldrake Seamount

- Vogel Seamount

2.29.4   See also

- New England hotspot

- Corner Rise Seamounts

- Seewarte Seamounts

2.29.5   References

[1] "Yale Peabody Museum: Invertebrate Zoology: Deep Sea Fauna from New England Seamounts" . *Yale Environmental News.* Yale University. 2004. Retrieved 2007-07-31.

[2] Ivar Babb (2005). "The New England Seamounts" . *National Oceanic and Atmospheric Administration.* U.S. Department of Commerce. Retrieved 2007-07-31.

[3] "Geological Origin of the New England Seamount Chain". *National Oceanic and Atmospheric Administration.* U.S. Department of Commerce. 2005. Retrieved 2007-07-31.

[4] Susan Mills (2005). "Seamount Coral Communities" . *National Oceanic and Atmospheric Administration.* U.S. Department of Commerce. Retrieved 2007-07-31.

[5] Petit, Charles (2004-08-08). "Denizens of the deep: In obscure marine ecosystems, clues to the origins of life" . *U.S. News & World Report.* Retrieved 2007-07-31.

2.29.6   External links

- Overview of Studies of NW Atlantic Seamounts for the ISA. International Seabed Authority.

2.30   Newfoundland Ridge

The **Newfoundland Ridge** is an ocean ridge in the northern Atlantic Ocean, located on the east coast of Canada. It was the site for major volcanic activity in the Barremian–Aptian period.[*][1]

2.30.1   See also

- Volcanism of Canada

- Volcanism of Eastern Canada

2.30.2   References

[1]

Coordinates: 40°4′55.5″N 48°0′0″W / 40.082083°N 48.00000°W

2.31   Newfoundland Seamounts

The **Newfoundland Seamounts** are a group of seamounts offshore of Eastern Canada in the northern Atlantic Ocean. Named for the island of Newfoundland, this group of seamounts formed during the Cretaceous period and are poorly studied.[*][1]

The Newfoundland Seamounts appear to have formed as a result of the North American Plate passing over the Azores hotspot. Scruncheon Seamount in the middle of the chain has given an isotopic date of 97.7 ± 1.5 million years for the Newfoundland Seamounts. This indicates that the Newfoundland Seamounts were volcanically active in the earliest Cenomanian stage.[*][2]

2.31.1   Seamounts

The Newfoundland Seamounts include:

- Shredder Seamount

- Scruncheon Seamount

2.31.2   See also

- Volcanism of Canada

- Volcanism of Eastern Canada

- List of volcanoes in Canada

2.31.3   References

[1] Paleocene volcanic sand provenance

[2] Roberts, David G.; Bally, A.W. (2012). *Regional Geology and Tectonics: Phanerozoic Passive Margins, Cratonic Basins and Global Tectonic Maps.* Elsevier. p. 357. ISBN 978-0-444-56357-6.

Coordinates: 43°41′27.9″N 45°24′15″W / 43.691083°N 45.40417°W

2.32   Panulirus Seamount

The **Panulirus Seamount** is a seamount in the Atlantic Ocean. It is part of the New England Seamount chain, which was active more than 100 million years ago. It was formed when the North American Plate moved over the New England hotspot.*[1]

2.32.1   References

[1] "Geological Origin of the New England Seamount Chain". *National Oceanic and Atmospheric Administration*. U.S. Department of Commerce. 2005. Retrieved 2007-09-16.

2.33   Physalia Seamount

The **Physalia Seamount** is a seamount in the Atlantic Ocean. It is part of the New England Seamount chain, which was active more than 100 million years ago. It was formed when the North American Plate moved over the New England hotspot.*[1]

2.33.1   References

[1] "Geological Origin of the New England Seamount Chain". *National Oceanic and Atmospheric Administration*. U.S. Department of Commerce. 2005. Retrieved 2007-09-16.

2.34   Picket Seamount

The **Picket Seamount** is a seamount in the Atlantic Ocean. It is part of the New England Seamount chain, which was active more than 100 million years ago. It was formed when the North American Plate moved over the New England hotspot.*[1]

2.34.1   References

[1] "Geological Origin of the New England Seamount Chain". *National Oceanic and Atmospheric Administration*. U.S. Department of Commerce. 2005. Retrieved 2007-09-16.

2.35   Protector Shoal

Protector Shoal is a submarine volcano, also called seamount, which rises gently from an ocean depth of 3,900 feet (1,200 m) to about 89 feet (27 m) below sea level approximately 31 miles (50 km) NW of Zavodovski Island in the South Sandwich Islands chain. The last eruption occurred during March 1962. Protector Shoal is the only volcano in the arc that has erupted rhyolite pumice.

2.35.1   See also

- List of volcanoes in South Sandwich Islands

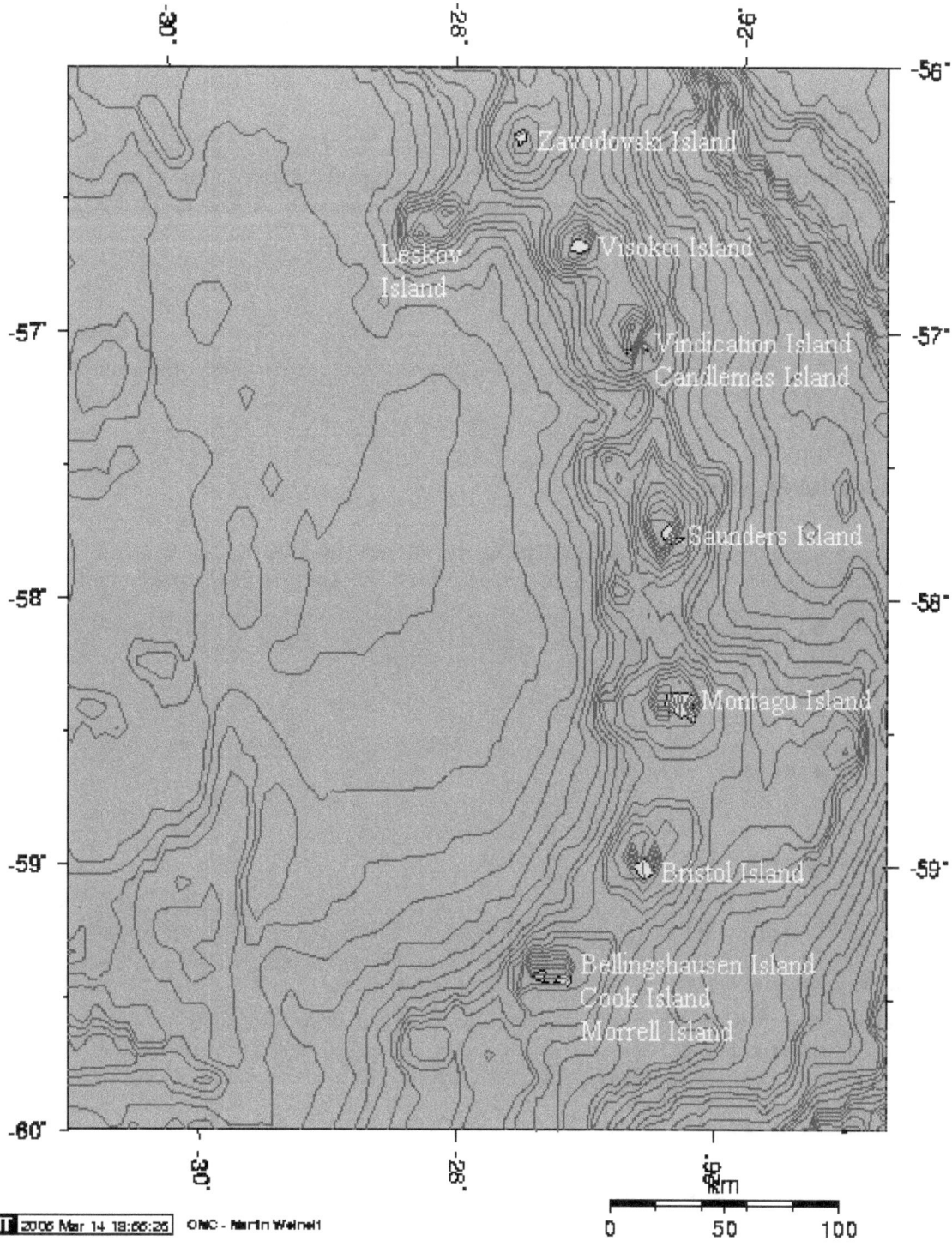

South Sandwich Islands

2.35.2   References

- LeMasurier, W. E.; Thomson, J. W. (eds.) (1990). *Volcanoes of the Antarctic Plate and Southern Oceans*. American Geophysical Union. p. 512 pp. ISBN 0-87590-172-7.

- "Protector Shoal" . *Global Volcanism Program*. Smithsonian Institution. Retrieved 2009-04-28.

- "Protector Shoal". *Volcano World*. Oregon State University. Retrieved 2009-04-28.

2.36 Rehoboth Seamount

The **Rehoboth Seamount** is a seamount in the Atlantic Ocean. It is part of the New England Seamount chain, which was active more than 100 million years ago. It was formed when the North American Plate moved over the New England hotspot.*[1]

2.36.1 References

[1] "Geological Origin of the New England Seamount Chain". *National Oceanic and Atmospheric Administration*. U.S. Department of Commerce. 2005. Retrieved 2007-09-16.

2.37 Retriever Seamount

The **Retriever Seamount** is a seamount in the Atlantic Ocean. It is part of the New England Seamount chain, which was active more than 100 million years ago. It was formed when the North American Plate moved over the New England hotspot.*[1]

2.37.1 References

[1] "Geological Origin of the New England Seamount Chain". *National Oceanic and Atmospheric Administration*. U.S. Department of Commerce. 2005. Retrieved 2007-09-16.

2.38 Rosemary Bank

Coordinates: 59°25′N 10°15′W / 59.417°N 10.250°W **Rosemary Bank** is a seamount approximately 120 kilometres west of Scotland, located in the Rockall Trough, in the northeast Atlantic. It was discovered in 1930 by the survey vessel HMS Rosemary, from which it takes its name.

The feature originated about 70 million years ago, as a result of volcanic activity. Rosemary Bank rises to approximately 2000 metres, its highest point being 300 metres below sea-level. The sea bed immediately surrounding it is approximately 2300 metres below sea-level. Around its base lies a thin "moat", where the sea-bottom is at a lower depth than the surrounding terrain.

2.39 San Pablo Seamount

The **San Pablo Seamount** is a seamount in the Atlantic Ocean. It is part of the New England Seamount chain, which was active more than 100 million years ago. It was formed when the North American Plate moved over the New England hotspot.*[1]

2.39.1 References

[1] "Geological Origin of the New England Seamount Chain". *National Oceanic and Atmospheric Administration*. U.S. Department of Commerce. 2005. Retrieved 2007-09-16.

2.40   Sedlo Seamount

Sedlo Seamount is an isolated seamount and underwater volcano located in the Northeast Atlantic, 180 mi (290 km) northeast of Graciosa Island. It has an elongate structure, roughly 75 by 30 km (47 by 19 mi). The summit is flat with three peaks. Sedlo Seamount sits on the ocean floor 3,000 m (9,843 ft) deep, and rises to within 660 m (2,165 ft) of the surface. Sedlo seamount has a tablemount structure, indicating that the peak of the seamount had once been above the water, but has since been ground down by persistent erosion to its current height. The seamount stands within the Exclusive Economic Zone of the Azores.*[1]

From 2002 to 2005, Sedlo Seamount was the target of a focused multidisciplinary study by the EU (titled OASIS), much of the research of which was published in 2009.

Complex hydrographical patterns with anticyclones and Taylor columns cause water flow around the summit. Water eddies tend to disrupt this flow. A bottom trawling experiment conducted during research brought up large orange roughy (*Hoplostethus atlanticus*) aggregations, as well as bycatch of benthic fauna including sponges, gorgonians, and scleractinian corals.*[1]

European Commission has enacted rules that protect Sedlo from bottom trawling, gillnets, and trammel nets. In 2007, Portugal proposed for Sedlo's inclusion in the OSPAR series of Marine Protected Areas. The motion was accepted in 2008, and a management plan for the MPA was being drafted. The MPA protects an area 62 by 65 km (39 by 40 mi).*[1]

2.40.1   See also

- Jasper Seamount

- Graveyard Seamounts

- Mud volcano

- Muirfield Seamount

- South Chamorro Seamount

2.40.2   References

[1] Ricardo S. Santos, Fernando Tempera, Gui Menezes, Filipe Porteiro, and Telmo Morato. "Spotlight 12: Sedlo Seamount" (PDF). *Oceanography*. Seamounts Special Issue (Oceanography Society) **23** (1). Retrieved 28 July 2010.

2.41   Seewarte Seamounts

The **Seewarte Seamounts**, also known as the **Atlantis-Great Meteor Seamount Chain** and the **Atlantis-Plato-Cruiser-Great Meteor Seamount Group**, is a north-south trending group of extinct submarine volcanoes in the northern Atlantic Ocean south-southeast of the Corner Rise Seamounts.

The Seewarte Seamounts have been interpreted to have formed as a result of the African Plate traveling over the New England hotspot.*[1]

2.41.1   Seamounts

The Seewarte Seamounts include:

- Closs Seamount

- Little Meteor Seamount

- Great Meteor Seamount

- Hyères Seamount

- Irving Seamount

- Cruiser Tablemount

- Plato Seamount

- Atlantis Seamount

- Tyro Seamount

2.41.2 See also

- Corner Rise Seamounts

- New England Seamount chain

2.41.3 References

[1] Geological Origin of the New England Seamount Chain

2.42 Sheldrake Seamount

The **Sheldrake Seamount** is a seamount in the Atlantic Ocean. It is part of the New England Seamount chain, which was active more than 100 million years ago. It was formed when the North American Plate moved over the New England hotspot.*[1]

2.42.1 References

[1] "Geological Origin of the New England Seamount Chain". *National Oceanic and Atmospheric Administration*. U.S. Department of Commerce. 2005. Retrieved 2007-09-16.

2.43 St. Helena Seamount chain

The **St. Helena Seamount chain**, also known as the **St. Helena Seamounts**, is an underwater chain of seamounts in the southern Atlantic Ocean. The chain has been formed by the movement of the African Plate over the Saint Helena hotspot.*[1]

2.43.1 References

[1] Plates, Plumes, and Paradigms

Coordinates: 15°44′29.3″S 6°25′52.5″W / 15.741472°S 6.431250°W

2.44   Vogel Seamount

The **Vogel Seamount** is a seamount in the Atlantic Ocean. It is part of the New England Seamount chain, which was active more than 100 million years ago. It was formed when the North American Plate moved over the New England hotspot.*[1]

2.44.1   References

[1] "Geological Origin of the New England Seamount Chain". *National Oceanic and Atmospheric Administration.* U.S. Department of Commerce. 2005. Retrieved 2007-09-16.

Chapter 3

Guyots

3.1   Abbott Seamount

Abbott Seamount is a seamount lying within the Hawaiian-Emperor seamount chain in the northern Pacific Ocean. It erupted 36-40 million years ago.[*][1]

3.1.1   See also

- List of volcanoes in the Hawaiian – Emperor seamount chain

3.1.2   References

[1] Abbott Seamount - John Search

3.2   Banc Capel

Banc Capel is a guyot, or flat-topped underwater volcano, in the Coral Sea.

3.2.1   Description

Banc Capel is a guyot – a former atoll with steep sides and a flat top – and is swept by strong currents. There are no sandy or muddy substrates, the surface being occupied by rocks or gravel scree.[*][2]

3.2.2   Biodiversity

Banc Capel is inhabited by species including *Nassarius alabasteroides* and *Laurentaeglyphea*.[*][3][*][4][*][5][*][6][*][2] It is dominated by sponges, including the genus *Phloedictyon* and gorgonians. Other decapods found in the same trawls including the slipper lobster *Ibacus brucei*, the crab *Randallia* and swimming crabs.[*][2]

3.2.3   Notes and references

[1]   "GeoHack". Retrieved 2014-09-25.

[2] Bertrand Richer de Forges (2006). "Découverte en mer du Corail d'une deuxième espèce de glyphéide (Crustacea, Decapoda, Glypheoidea)" (PDF). *Zoosystema* **28** (1): 17–28.

[3] Kool (2009). Miscellanea Malacologica 3 (5) : 97-100. World Register of Marine Species, Retrieved 18 April 2010.

[4] *Nassarius alabasteroides* Kool, 2009. Retrieved through: World Register of Marine Species on 18 April 2010.

[5] *Nassarius alabasteroides Kool, 2009*, Muséum national d'histoire naturelle, 2009

[6] T. Y. Chan, M. Butler, A. MacDiarmid, A. Cockcroft & R. Wahle (2011). "*Laurentaeglyphea neocaledonica*". *IUCN Red List of Threatened Species. Version 2011.2.* International Union for Conservation of Nature. Retrieved January 7, 2012.

3.3   Bowie Seamount

Bowie Seamount is a large submarine volcano in the northeastern Pacific Ocean, located 180 km (110 mi) west of Haida Gwaii, British Columbia, Canada.

The seamount is named after William Bowie of the Coast & Geodetic Survey.[*][3]

The volcano has a flat-topped summit (thus making it a guyot) rising about 3,000 m (10,000 ft) above the seabed, to 24 m (79 ft) below sea level.[*][1] The seamount lies at the southern end of a long underwater volcanic mountain range called the Pratt-Welker or Kodiak-Bowie Seamount chain, stretching from the Aleutian Trench in the north almost to the Queen Charlotte Islands in the south.[*][1]

Bowie Seamount lies on the Pacific Plate, a large segment of the Earth's surface which moves in a northwestern direction under the Pacific Ocean. Its northern and eastern flanks are surrounded by neighboring submarine volcanoes; Hodgkins Seamount on its northern flank and Graham Seamount on its eastern flank.

3.3.1   Geology

Structure

Seamounts are volcanic mountains which rise from the seafloor. The unlimited supply of water surrounding these volcanoes can cause them to behave differently from volcanoes on land. The lava emitted in eruptions at Bowie Seamount is made of basalt, a common gray to black or dark brown volcanic rock low in silica content (the lava is mafic). When basaltic lava makes contact with the cold sea water, it may cool very rapidly to form pillow lava, through which the hot lava breaks to form another pillow. Pillow lava is typically fine-grained, due to rapid cooling, with a glassy crust, and has radial jointing.[*][4]

With a height of at least 3,000 m (10,000 ft) and rising to within only 24 m (79 ft) of the sea surface, Bowie Seamount is the shallowest submarine volcano on the British Columbia Coast, as well as in Canadian waters, and one of the shallowest submarine volcanoes in the northeast Pacific Ocean.[*][1][*][2] Most seamounts are found hundreds to thousands of metres below sea level, and are therefore considered to be within the deep sea. In contrast, if Bowie Seamount were on land it would be about 600 m (2,000 ft) higher than Whistler Mountain in southwestern British Columbia and 800 m (2,600 ft) lower than Mount Robson, the highest mountain in the Canadian portion of the Rocky Mountains.[*][2]

Bowie Seamount is about 55 km (34 mi) long and 24 km (15 mi) wide.[*][2] Its flat-topped summit is made of weakly consolidated tephra and consists of two terraces.[*][2][*][5] The lowest terrace is about 230 m (750 ft) below sea level while the highest is about 80 m (260 ft) below sea level, but contains steep-sided secondary summits that rise to within 25 m (82 ft) below sea level. From a physical perspective, the effective size of the submarine volcano is possibly a lot greater than its mass alone would suggest. The effects of other submarine volcanoes along the Pacific Northwest, including Cobb Seamount off the coast of Washington, can be noticed in the composition and abundance of the tiny floating organisms called plankton up to 30 km (20 mi) away from the seamount summit. Because of its similar size, Bowie Seamount most likely has a similar effect on its adjacent waters.[*][2]

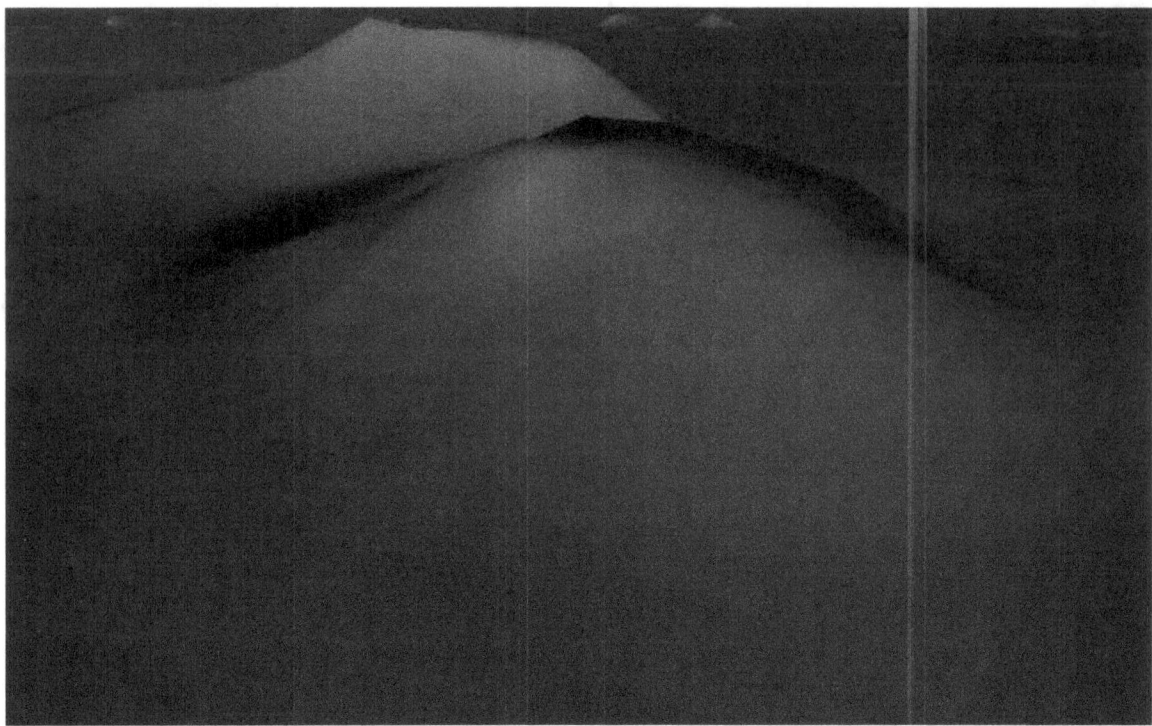

3-D depiction of Hodgkins Seamount with Bowie Seamount in the background

Eruptive history

Bowie Seamount was formed by submarine eruptions along fissures in the seabed throughout the last glacial, or "Wisconsinian", period, which began about 110,000 years ago and ended between 10,000 and 15,000 years ago. While most submarine volcanoes in the Pacific Ocean are more than one million years old, Bowie Seamount is relatively quite young. Its base was formed less than one million years ago but its summit shows evidence of volcanic activity as recently as 18,000 years ago.[2] This is very recent in geological terms, suggesting the volcano may yet have some ongoing volcanic activity.[1]

Close to Bowie's submerged summit, former coastlines cut by wave actions and beach deposits show that the submarine volcano would once have stood above sea level, as either a single volcanic island or as a small cluster of shoals that would have been volcanically active. Sea levels during the last glacial period, when Bowie Seamount was formed, were at least 100 m (300 ft) lower than they are today.[1][2]

Origins

The origin of the volcanism that produced Bowie Seamount is not without controversy. Geological studies indicate that the Kodiak-Bowie Seamount chain may have formed above a center of upwelling magma called a mantle plume. The seamounts comprising the Kodiak-Bowie Seamount chain would be formed above the mantle plume and carried away from the mantle plume's magmatic source as the Pacific Plate moves in a northwesterly direction towards the Aleutian Trench, along the southern coastline of Alaska.[6]

The volcanic rocks which make up some of the seamounts in the Kodiak-Bowie Seamount chain are unusual in that they have an acid-neutralizing chemical substance like typical ocean-island basalts but a low percentage of strontium as found at mid-ocean ridge basalts. However, the strontium-bearing volcanic rocks comprising Bowie Seamount also contain lead. Therefore, the magma mixtures that formed Bowie Seamount seem to have originated from varying degrees of partial melting of a depleted source in the Earth's mantle and basalts which had distinctly high lead isotopic ratios. Estimates during geological studies indicate that the abundance of the depleted-source component ranges from 60 to 80 percent.[6]

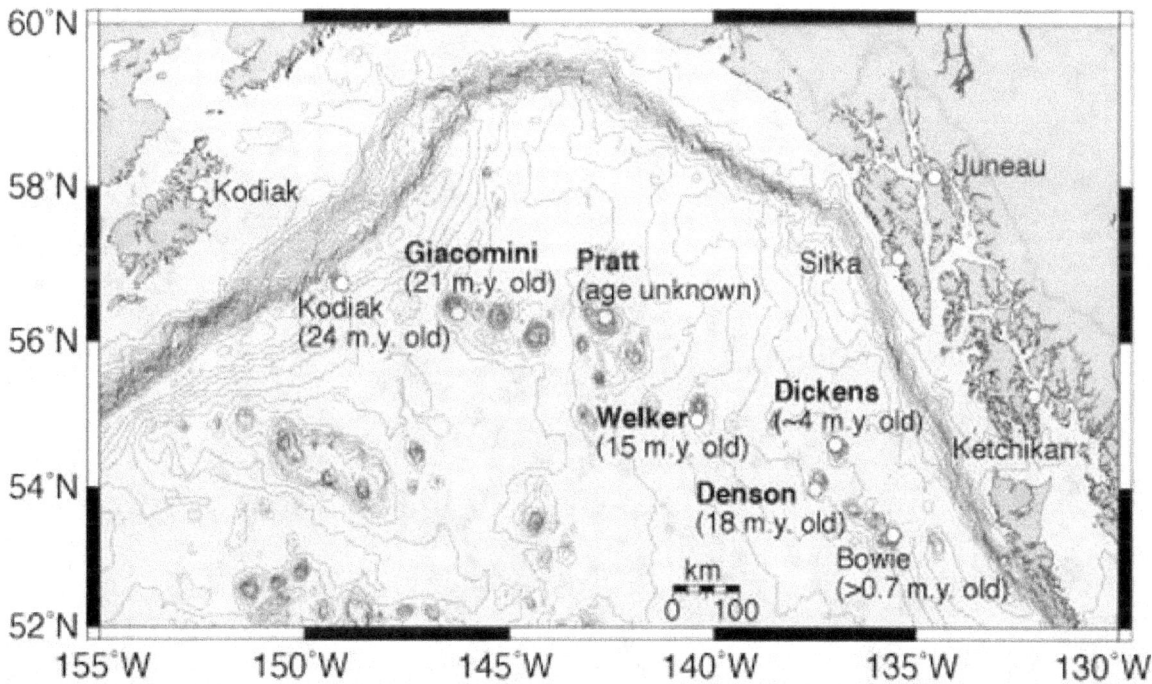

Map of the Kodiak-Bowie Seamount chain

Some aspects of the origin of the Kodiak-Bowie Seamount chain remain uncertain. The volcanic rocks found at the Tuzo Wilson Seamounts south of Bowie are fresh glassy pillow basalts of recent age, as would be expected if these seamounts are located above or close to a mantle plume south of the Queen Charlotte Islands. However, the origin of Bowie Seamount is less certain because even though the seafloor which Bowie lies on formed 16 million years ago during the late Miocene period, Bowie's summit shows evidence of recent volcanic activity. If Bowie Seamount formed above a mantle plume at the site presently occupied by the Tuzo Wilson Seamounts, it has been displaced from its magma source by about 625 km (390 mi) at a rate of about 4 cm (2 in) per year. The geologic history of Bowie Seamount is consistent with its flat-topped eroded summit, but the source for Bowie's recent volcanic activity remains uncertain.*[6] Still others, such as Dickens Seamount and Pratt Seamount further north of Bowie Seamount, fall a little to the side of the chain's expected trend.*[7] Another hypothesized origin of some or all seamounts in the Kodiak-Bowie Seamount chain is that they formed on top of the Explorer Ridge, a divergent tectonic plate boundary west of Vancouver Island, and have been displaced from it by seafloor spreading.*[6]

Although some of the seamounts in the Kodiak-Bowie Seamount chain appear to follow the expected age progression for a mantle plume trail, others, such as Denson Seamount, are older than that hypothesis would suggest.*[7] As a result, the Kodiak-Bowie Seamount chain has also been proposed by geoscientists to be a mix of ridge and mantle plume volcanism.

3.3.2 Biology

Bowie Seamount supports a biologically rich area with a vigorous ecosystem. Studies have recorded high densities of crab, sea stars, sea anemones, sponges, squid, octopus, rockfish, halibut and sablefish. Eight species of marine mammal have been found in the Bowie Seamount area, including Steller sea lions, orca, humpback and sperm whales, along with 16 varieties of seabirds.*[8] This has made Bowie Seamount a rare habitat in the northeast Pacific Ocean and one of the most biologically rich submarine volcanoes on Earth.*[1]*[9] The rich marine life is due to the intense food supply of microscopic animals and plants, including phytoplankton and zooplankton.*[10]

Haida Heritage Centre at Kaay Llnagaay where the Bowie Seamount Marine Protected Area was announced

Bowie Seamount Marine Protected Area

Because of its biological richness, Bowie Seamount was designated as Canada's seventh Marine Protected Area on April 19, 2008 under the Oceans Act and has been described as an "Oceanic Oasis".*[8] The announcement was made by federal Fisheries Minister Loyola Hearn and Guujaaw, President of the Council of the Haida Nation, in Skidegate on the Queen Charlotte Islands, also called Haida Gwaii. During the announcement, Natural Resources Minister Gary Lunn said: "Bowie Seamount is an oceanic oasis in the deep sea, a rare and ecologically rich marine area, and our government is proud to take action to ensure it is protected. By working in partnership with the Council of the Haida Nation and groups like the World Wildlife Fund-Canada, we are ensuring this unique treasure is preserved for future generations." *[9] It measures about 118 km (73 mi) long and 80 km (50 mi) wide, totaling an area of 6,131 km^2 (2,367 sq mi).*[10] This is the northernmost of the two Marine Protected Areas on the British Columbia Coast; the southernmost is the Endeavour Hydrothermal Vents, an active hydrothermal vent zone of the Juan de Fuca Ridge 250 km (160 mi) southwest of Vancouver Island.*[11] The Bowie Seamount Marine Protected Area also includes Peirce Seamount (also called Davidson Seamount) and Hodgkins Seamount.*[9]

3.3.3   Diving explorations and studies

The shallow depth of Bowie Seamount makes it the only underwater mountain off the British Columbia Coast easily reached using scuba diving equipment. In March 1969, dives were made at the submarine volcano by Canadian Forces Maritime Command divers from the CSS Parizeau during a new study for device package placement. Two dives were made to the summit where monochrome photographs were taken to establish the environment of Bowie's base and some biological tests were gathered to detect possible harmful organisms, including plants, animals, or bacteria. These specimens were identified at the Pacific Biological Station in Nanaimo, creating a list of eleven varieties of sea bottom invertebrates.[5]

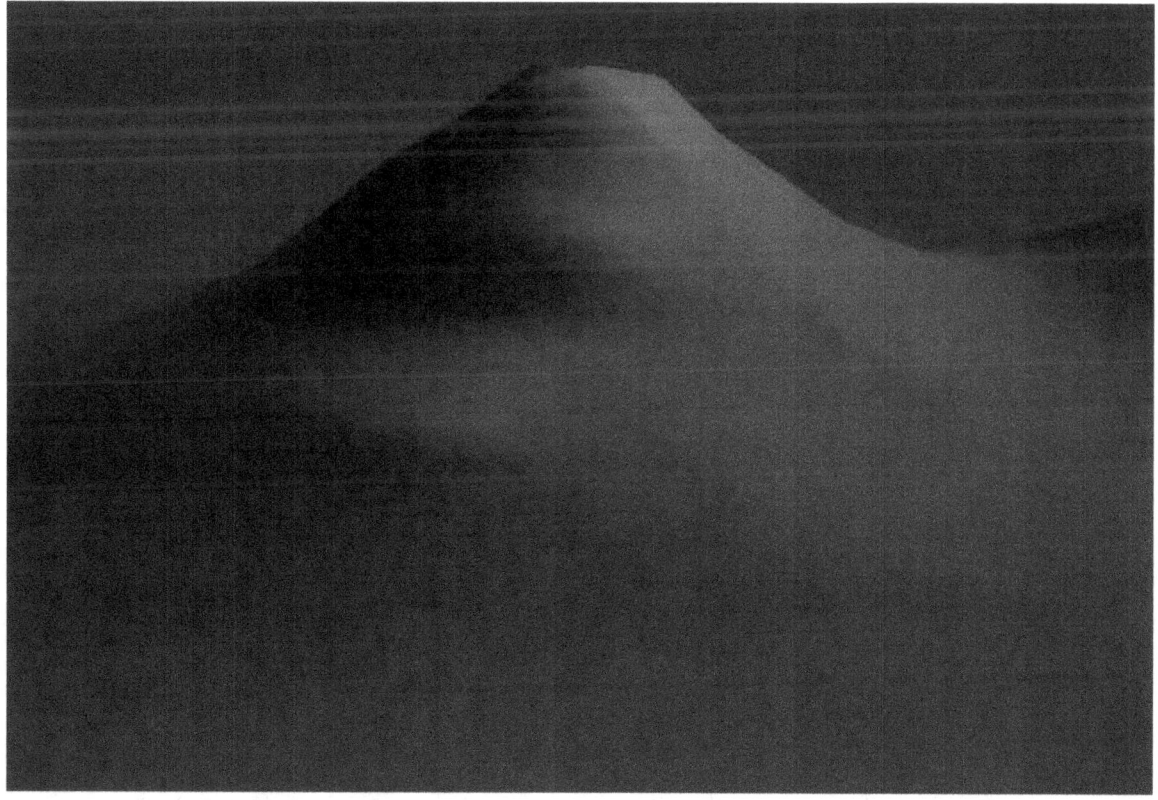

Eastern flank of Bowie Seamount

In August 1969, Canadian Forces Maritime Command divers made more dives during scientific studies by the Fisheries Research Board of Canada. They recognized the existence of very thick groups of rockfish floating on top of Bowie's flat-topped summit and a variety of bottom life. A number of monochrome photographs were taken and a few seaweeds were gathered for documents, but no species record was created for other types of oceanic life around Bowie Seamount.[5]

In November 1996, a release of the National Geographic Magazine comprised a piece of writing titled "Realm of the Seamount", illustrating dives made at Bowie Seamount by two diving explorers named Bill Curtsinger and Eric Hiner. They explored Bowie Seamount down to depths of 50 m (160 ft) using scuba diving equipment while the slopes of the seamount down to 150 m (490 ft). Images photographed by the two diving explorers featured one of Bowie's rugged peaks covered with thick seaweeds and multicoloured sea bottom invertebrates. Thick groups of young rockfish were detected on Bowie's steep flanks.[5]

Scientist Bill Austin of Khoyatan Marine Lab in the Northeast Pacific looked over a video film earned throughout the National Geographic dives to verify the flora and fauna of the sea bottom surrounding Bowie Seamount. From the video film, Austin recognized some of the most noticeable invertebrates and noted that a few species more regularly occurring between high-tide and low-tide marks and shallow environments were found deeper than might normally be expected, and were bigger than normal.[5]

A team of five divers, led by photographer/videographer Neil McDaniel, visited the seamount August 3–5, 2003 and conducted a biological and photographic survey of the summit down to depths of about 40 m (130 ft). A total of 18 taxa of algae, 83 taxa of conspicuous invertebrates and 12 taxa of fishes were documented, approximately 180 underwater still photographs were taken and approximately 90 minutes of digital video were recorded. Of particular note were the dense schools of rockfish hovering over the summit and numerous curious prowfish.[5]

3.3.4   Indigenous people

To the Haida Nation, the indigenous people who played a key role to establish the Bowie Seamount Marine Protected Area, the submarine volcano is called *Sgaan Kinghlas*. In their language it means "Supernatural Being Looking Outward" .[*][9]

This seamount has long been recognized by the Haida Nation as a special place. Guujaaw, President of the Council of the Haida Nation, has said: "Sgaan Kinghlas represents a shift in recognizing the need for respect and care for the Earth. This is a very significant turning point in reversing the trends that have been leading to the depletion of life in the sea." [*][9]

3.3.5   Marine hazard

Given its shallow depth, Bowie Seamount is a potential marine hazard because of the strong storms that strike the British Columbia Coast during winter. Waves have been recorded with heights of more than 20 m (70 ft),[*][2] enough to expose the summit and cause devastation to any vessel transitting the area. For this reason, Bowie Seamount is recognized as a hazard to navigation and is avoided by shipping vessels.[*][12]

3.3.6   See also

- Volcanism of Canada

- Volcanism of Western Canada

- List of volcanoes in Canada

3.3.7   References

[1] "Bowie Seamount Marine Protected Area Management Plan" (PDF). Fisheries and Oceans Canada. August 2001. Archived from the original (PDF) on 2009-03-24. Retrieved October 27, 2008.

[2] "The Bowie Seamount Area" (PDF). John F. Dower and Frances J. Fee. February 1999. Archived from the original (PDF) on 2009-03-24. Retrieved October 27, 2008.

[3] "Undersea Features History" . GEOnet Names Server. Retrieved 2012-03-97. Check date values in: |access-date= (help)

[4] "Submarine Volcanoes, Vents, Ridges, and Eruptions" . USGS. March 13, 2005. Retrieved October 27, 2008.

[5] "Biological Observations at Bowie Seamount: August 3-5, 2003" (PDF). N. McDaniel, D. Swanston, R. Haight, D. Reid and G. Grant. October 22, 2003. Archived from the original (PDF) on 2009-03-24. Retrieved November 6, 2008.

[6] Faure, Gunter (2001). *Origin of Igneous Rocks: The Isotopic Evidence*. Springer. pp. 66, 67. ISBN 978-3-540-67772-7.

[7] "Seamounts in the Eastern Gulf of Alaska: A Volcanic Hotspot with a Twist?". National Oceanic and Atmospheric Administration. July 12, 2005. Retrieved October 27, 2008.

[8] "Bowie Seamount Marine Protected Area" . Fisheries and Oceans Canada. May 25, 2006. Archived from the original on 2009-03-10. Retrieved October 27, 2008.

[9] "Bowie Seamount Designated as Canada's Seventh Marine Protected Area" . Fisheries and Oceans Canada. April 19, 2008. Retrieved October 27, 2008.

[10] "Designation of Bowie Seamount as a Marine Protected Area" . Fisheries and Oceans Canada. October 11, 2008. Retrieved October 30, 2008.

[11] "Marine Protected Areas" . Fisheries and Oceans Canada. March 7, 2008. Retrieved November 7, 2008.

[12] "Record of Meeting" . Fisheries and Oceans Canada. February 19, 2002. Archived from the original on June 20, 2004. Retrieved October 27, 2008.

3.3.8   External links

- Bowie Seamount protected on B.C. Coast

- Bowie Seamount Marine Protected Area

3.4   Colahan Seamount

Colahan Seamount is a seamount lying within the Hawaiian-Emperor seamount chain in the northern Pacific Ocean. It erupted 37-40 million years ago.[*][1]

3.4.1   See also

- List of volcanoes in the Hawaiian – Emperor seamount chain

3.4.2   References

[1] Colahan Seamount - John Search

3.5   Daikakuji Guyot

Daikakuji Seamount is a seamount (underwater volcano) and the southwesternmost volcanic feature in the Hawaiian Emperor chain bend area.

3.5.1   Geology

The seamount is very close to the "V"-shaped bend in the Hawaiian-Emperor seamount chain, and thus would be useful in understanding the exact age of the bend. Although few dredge samples are available, they have all been reliably dated at 43 million years, during the Eocene epoch of the Paleogene period.

During the cruise SO112 of the *R/V SONNE*, high resolution bathymetric mapping was conducted, showing that Daikakuji is nearly 30 km (19 mi) in diameter and nearly 4,000 m (13,123 ft) in height, with a summit lying 1,000 m (3,281 ft) underwater.

Because of its flat capped top, Daikakuji is considered a guyot. A smaller, younger, secondary guyot just east of the main mass overlaps its slope. The western site suffered a large collapse sometime in its history, evident by a large slump, that likely carried away a significant part of the volcano's caldera.

Daikakuji Seamount has some well developed rift zones oriented towards the Emperor portion of the chain, whereas the younger, secondary cone has rift flanks in the direction of the Hawaiian ridge.

3.5.2   See also

- List of volcanoes in the Hawaiian – Emperor seamount chain

3.5.3   References

[1] B. C. Kerr; D. W. Scholl; and S. L. Klemperer (12 July 2005). "Seismic stratigraphy of Detroit Seamount, Hawaiian-Emporer Seamount chain" (PDF). *Scientific Publication*. Stanford University. Retrieved 2009-04-03.

[2] "DRILLING STRATEGY". OCean Drilling Program. Retrieved 2009-04-04.

3.6 Detroit Seamount

Detroit Seamount, which was formed around 76 million years ago, is one of the oldest seamounts of the Hawaiian-Emperor seamount chain (Meiji Seamount is the oldest, at 82 million years). It lies near the northernmost end of the chain and is south of Aleutian Islands (near Russia),[1] at 51°28.80′N 167°36′E / 51.48000°N 167.600°ECoordinates: 51°28.80′N 167°36′E / 51.48000°N 167.600°E[2] It is a seamount in the chain, located north of the hinge of the "V" in the image at right.[1]

Detroit Seamount is one of the few seamounts to break the naming scheme of the Emperor seamounts, which are named mostly after emperors or empresses of the Kofun period of Japanese history. It is instead named after the light cruiser *USS Detroit.*[5]

The Detroit Seamount is as big as the island of Hawaii.[1]

3.6.1 Mapping

The seamount was initially mapped by the GLORIA program of the USGS, and in far more detail in 2001 by leg 197 of the Ocean Drilling Program (ODP). 2001 marked a two-month excursion aboard the research vessel *JOIDES Resolution* to collect samples of lava flows from four submerged volcanoes, among them Detroit Seamount, which was drilled twice. The expedition was funded by the Ocean Drilling Program, an international research effort designed to study the world's seafloors, and the drill sites were numbers 1203 through 1206. The project drilled Detroit, Nintoku, and Koko seamounts, all in the far northwest of the chain.[1] Detroit Seamount was drilled twice (numbered 1203 and 1204), on the summit and on one its secondary cones; care was taken to put the locations away from major fault lines or other geological features that would otherwise invalidate or bias the results.[1]

In 2005 it underwent a detailed geological analysis by scientists from Stanford University.[1]

3.6.2 Geology

After its initial formation 51 million years ago, the volcano was active for 25 million years. Parts of the volcano appear to be older than the oldest volcano in the chain, Meiji Seamount. The 2005 analysis found that the volcano had been active throughout much of the Eocene (circa 52-34 million years ago), and that activity may have extended into the Oligocene (under 34 million years ago).[1] The large difference between the youngest and oldest lavas provides evidence that the Hawaii hotspot migrated far more slowly than it does today; for example, Kohala volcano (the oldest volcano of Hawaii island) first emerged from the sea 500,000 years ago, and last erupted 120,000 years ago, a period of only 380,000 years in comparison to Detroit's 18 million or more years of volcanic activity.[1] The large age difference (51 vs. 34 million years) between the submarine preshield stages and the post-shield rejuvenated stage seems to indicate that volcanoes in the chain can erupt again long after they are believed to be extinct. The volcano is known to have erupted intermediately in an underwater and shallow-water environment.[1]

Detroit Seamount has a wide (100,000 square kilometers) base and rises from the bottom of the abyssal plain to a depth of approximately 1,550 m (5,085 ft); in fact, it is as wide as Hawaii island at the head of the chain. The width of the seamount, as well as the extremely gentle slope, which is very shallow even for a Hawaiian shield volcano, seem to show that the seamount suffered a catastrophic collapse sometime in its history; such a collapse is a relatively common event in the growth of Hawaiian volcanoes, caused when the volcanoes grow so fast that they destabilize.[1] A sequence of sediments 800 m (2,600 ft) to 900 m (3,000 ft) thick compose the volcano, in several layers. Some papers refer to only the shallowest part of the volcano as Detroit Seamount, and the rest of the seamount as the "Detroit Rise." The tallest volcanic cones of the seamount peak 1 km (0.62 mi) to 2 km (1.2 mi) above the rest of the seamount.[1]

3.6.3 Mantle of sediment

The seamount was thought to be covered in a cap of sediment, which was confirmed in 2005. All but the topmost cones of Detroit Seamount are capped in a thick layer of sediments, which were found to have drifted there from a direction due northwest. The drift that carried the sediments onto the volcano was named the "Meiji Drift," after the oldest volcano in the chain, Meiji Seamount, which was also in that direction. The drift is of Oligocene to Quaternary-era mud, deposited by ocean currents. The tallest parts of the seamount protrude above this "mud cap," which at its deepest is estimated to be 840 m (2,756 ft) thick. They formed 34 million years ago.[*][1]

A 2005 analysis of the results of the 2001 *JOIDES Resolution* excursion found the age, composition, structure, and history of growth for the seamount. The evaluation also focused on the strange cones that poked through the sedimentary layers. They were deposited onto the seamount before the Meiji Drift developed. Analysis put the latest date of their formation at 60 million years ago, 6 million years into the seamount's life.[*][1]

3.6.4   References

[1] Kerr, Bryan C.; Scholl, David W.; Klemperer, Simon L. (July 12, 2005). "Seismic stratigraphy of Detroit Seamount, Hawaiian–Emperor Seamount chain" (PDF). Stanford University. doi:10.1029/2004GC000705. Retrieved April 3, 2009.(registration required)

[2] Seamounts Catalog by EarthRef, a National Science Foundation Project accessed 3-1-09.

[3] "DRILLING STRATEGY". Ocean Drilling Program. Retrieved 2009-04-04.

[4] Regelous, M.; Hofmann, A.W.; Abouchami, W.; Galer, S.J.G. (2003). "Geochemistry of Lavas from the Emperor Seamounts, and the Geochemical Evolution of Hawaiian Magmatism from 85 to 42 Ma" (PDF). *Journal of Petrology* **44** (1): 113–140. doi:10.1093/petrology/44.1.113. Retrieved July 23, 2010.

[5] Calgue, David A.; Dalrymple, G. Brent; Greene, H. Gary; Wald, Donna; Kono, Masaru; Kroenke, Loren W. (1980). "40. Bathymetry of the Emperor Seamounts" (PDF). *Initial Reports of the Deep Sea Drilling Project* **55**. Washington, DC: US Government Printing Office. pp. 846–847. LCCN 74-603338. Retrieved April 25, 2012.

3.7   Hancock Seamount

Hancock Seamount is a seamount of the Hawaiian-Emperor seamount chain in the Pacific Ocean.

It was formed in the Eocene and Oligocene epochs of the Paleogene Period. The last eruption from Hancock Seamount is unknown.[*][1]

3.7.1   See also

- List of volcanoes in the Hawaiian – Emperor seamount chain

3.7.2   References

[1] Handcock Seamount - John Search

3.8   Hawaiian–Emperor seamount chain

The **Hawaiian–Emperor seamount chain** is a mostly undersea mountain range in the Pacific that reaches above sea level in Hawaii. It is composed of the Hawaiian ridge, consisting of the islands of the Hawaiian chain northwest to Kure Atoll, and the Emperor Seamounts: together they form a vast underwater mountain region of islands and intervening seamounts, atolls, shallows, banks and reefs along a line trending southeast to northwest beneath the northern Pacific Ocean. The seamount chain, containing over 80 identified undersea volcanoes, stretches over 5,800 kilometres (3,600 mi) from the Aleutian Trench in the far northwest Pacific to the Loʻihi seamount, the youngest volcano in the chain, which lies about 35 kilometres (22 mi) southeast of the Island of Hawaiʻi.

3.8.1 Regions

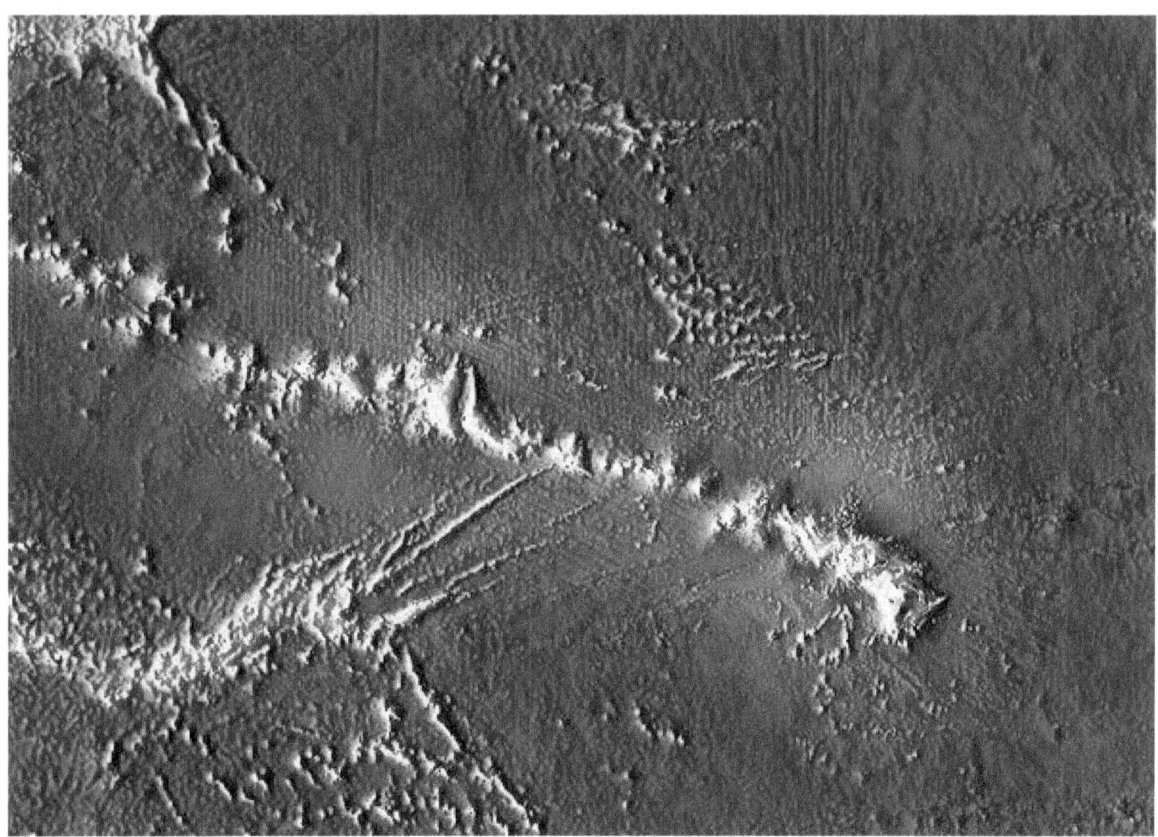

The Hawaiian-Emperor seamount chain, zoomed in on the current habitable islands

The chain can be divided into three subsections. The first, the Hawaiian archipelago (also known as the *Windward isles*), consists of the islands comprising the U.S. state of Hawaii. As it is the closest to the hotspot, this volcanically active region is the youngest part of the chain, with ages ranging from 400,000 years[1] to 5.1 million years.[2] The island of Hawai'i is composed of five volcanoes, of which three (Kilauea, Mauna Loa, and Hualalai) are still active. Lō'ihi Seamount continues to grow offshore, and is the only known volcano in the chain in the submarine pre-shield stage.[3]

The second part of the chain is composed of the Northwestern Hawaiian Islands, collectively referred to as the *Leeward isles*, the constituents of which are between 7.2 and 27.7 million years in age.[2] Erosion has long since overtaken volcanic activity at these islands, and most of them are atolls, atoll islands, and extinct islands. They contain many of the most northerly atolls in the world; one of them, Kure Atoll, is *the* northernmost atoll in the world.[4] On June 15, 2006, U.S. President George W. Bush issued a proclamation creating Papahānaumokuākea Marine National Monument under the Antiquities Act of 1906. The national monument, meant to protect the biodiversity of the Hawaiian isles,[n 1] encompasses all of the northern isles, and is one of the largest such protected areas in the world. The proclamation limits tourism to the area, and calls for a phase-out of fishing by 2011.[5]

The oldest and most heavily eroded part of the chain are the Emperor seamounts, which are 39[6] to 85 million years in age.[7] The Emperor and Hawaiian chains form an angle of about 120°. This bend was long attributed to a relatively sudden change of 60° in the direction of plate motion, but research conducted in 2003 suggests that it was the movement of the hotspot itself that caused the bend.[8] The issue is still currently under debate.[9] All of the volcanoes in this part of the chain have long since subsided below sea level, becoming seamounts and guyots (see also the seamount and guyot stages of Hawaiian volcanism). Many of the volcanoes are named after former emperors of Japan. The seamount chain extends to the West Pacific, and terminates at the Kuril–Kamchatka Trench, a subduction zone at the border of Russia.[10]

3.8.2 Formation

Main article: Evolution of Hawaiian volcanoes

The oldest age for the Emperor Seamounts is 81 million years, and comes from Detroit Seamount. However, Meiji Guyot, located to the north of Detroit Seamount, is likely somewhat older.

In 1963, geologist John Tuzo Wilson hypothesized the origins of the Hawaiian–Emperor seamount chain, explaining that they were created by a hotspot of volcanic activity that was essentially stationary as the Pacific tectonic plate drifted in a northwesterly direction, leaving a trail of increasingly eroded volcanic islands and seamounts in its wake. An otherwise inexplicable kink in the chain marks a shift in the movement of the Pacific plate some 47 million years ago, from a northward to a more northwesterly direction, and the kink has been presented in geology texts as an example of how a tectonic plate can shift direction comparatively suddenly. A look at the USGS map on the origin of the Hawaiian Islands[11] clearly shows this "spearpoint".

In a more recent study, Sharp and Clague (2006) interpret the bend as starting at about 50 million years ago. They also conclude that the bend formed from a "traditional" cause—a change in the direction of motion of the Pacific plate.

However, recent research shows that the hotspot itself may have moved with time. Some evidence comes from analysis of the orientation of the ancient magnetic field preserved by magnetite in ancient lava flows sampled at four seamounts (Tarduno et al., 2003): this evidence from paleomagnetism shows a more complex history than the commonly accepted view of a stationary hotspot. If the hotspot had remained above a fixed mantle plume during the past 80 million years, the latitude as recorded by the orientation of the ancient magnetic field preserved by magnetite (paleolatitude) should be constant for each sample; this should also signify original cooling at the same latitude as the current location of the Hawaiian hotspot. Instead of remaining constant, the paleolatitudes of the Emperor Seamounts show a change from north to south, with decreasing age. The paleomagnetic data from the seamounts of the Emperor chain suggest motion of the Hawaiian hotspot in Earth's mantle. Tarduno et al. (2009) have summarized evidence that the bend in the seamount chain may be caused by circulation patterns in the flowing solid mantle (mantle "wind") rather than a change in plate motion.

3.8.3 Aging

The chain has been produced by the movement of the ocean crust over the Hawaiʻi hotspot, an upwelling of hot rock from the Earth's mantle. As the oceanic crust moves the volcanoes farther away from their source of magma, their eruptions become less frequent and less powerful until they eventually cease to erupt altogether. At that point erosion of the volcano and subsidence of the seafloor cause the volcano to gradually diminish. As the volcano sinks and erodes, it first becomes an atoll island and then an atoll. Further subsidence causes the volcano to sink below the sea surface, becoming a seamount and/or a guyot.[3]

3.8.4 See also

- List of volcanoes in the Hawaiian – Emperor seamount chain
- Detroit Seamount
- Evolution of Hawaiian volcanoes
- Isostasy
- Kodiak–Bowie Seamount chain
- Meiji Seamount
- New England Seamount chain
- Oceanic trench
- Pacific-Kula Ridge

- Plate tectonics

- Timeline of the far future

3.8.5 Notes

[1] All of the islands in this part of the chain are administrated by Hawaii state, save for Midway Atoll, which is administrated by the U.S. Fish and Wildlife Service.

3.8.6 References

[1] Michael O. Garcia, Jackie Caplan-Auerbanch, Eric H. De Carlo, M.D. Kurz, N. Becker (September 20, 2005). "Geology, geochemistry and earthquake history of Lōʻihi Seamount, Hawaiʻi" (PDF). *Chemie der Erde - Geochemistry*. This is the pre-press version of a paper that was published on 2006-05-16 as "Geochemistry, and Earthquake History of Lōʻihi Seamount, Hawaiʻi's youngest volcano", in *Chemie der Erde – Geochemistry* (66) 2:81–108 (University of Hawaii – School of Ocean and Earth Science and Technology) **66**: 81–108. Bibcode:2006ChEG...66...81G. doi:10.1016/j.chemer.2005.09.002. Retrieved March 20, 2009. Pre-press version

[2] Rubin, Ken. "The Formation of the Hawaiian Islands". Hawaii Center for Vulcanology. Retrieved May 18, 2009.

[3] "Evolution of Hawaiian Volcanoes". Hawaiian Volcano Observatory (USGS). September 8, 1995. Retrieved March 7, 2009.

[4] "Kure Atoll". Public Broadcasting System – KQED. March 22, 2006. Retrieved June 13, 2009.

[5] Staff authors (June 15, 2006). "Bush creates new marine sanctuary". BBC News. Retrieved December 14, 2009.

[6] Sharp, W. D.; Clague, DA (2006). "50-Ma Initiation of Hawaiian-Emperor Bend Records Major Change in Pacific Plate Motion". *Science* **313** (5791): 1281–84. Bibcode:2006Sci...313.1281S. doi:10.1126/science.1128489. PMID 16946069.

[7] Regelous, M.; Hofmann, A.W.; Abouchami, W.; Galer, S.J.G. (2003). "Geochemistry of Lavas from the Emperor Seamounts, and the Geochemical Evolution of Hawaiian Magmatism from 85 to 42 Ma" (PDF). *Journal of Petrology* **44** (1): 113–140. doi:10.1093/petrology/44.1.113. Retrieved July 23, 2010.

[8] John Roach (August 14, 2003). "Hot Spot That Spawned Hawaii Was on the Move, Study Finds". National Geographic News. Retrieved March 9, 2009.

[9] Sharp *et al.*, 2006, *Initiation of the bend near Kimmei seamount about 50 million years ago (MA) was coincident with realignment of Pacific spreading centers and early magmatism in western Pacific arcs, consistent with formation of the bend by changed Pacific plate motion.*

[10] G. R. Foulger; Don L. Anderson. "The Emperor and Hawaiian Volcanic Chains: How well do they fit the plume hypothesis?". MantlePlumes.org. Retrieved April 1, 2009.

[11] "origin of the Hawaiian Islands". Pubs.usgs.gov. 2013-01-04. Retrieved 2013-01-12.

3.8.7 Further reading

- Tarduno, John A.; et al. (2003). "The Emperor Seamounts: Southward Motion of the Hawaiian Hotspot Plume in Earth's Mantle". *Science* **301** (5636): 1064–1069. Bibcode:2003Sci...301.1064T. doi:10.1126/science.1086442. PMID 12881572.

- Tarduno, John A.; et al. (2009). "The Bent Hawaiian-Emperor Hotspot Track: Inheriting the Mantle Wind". *Science* **324**: 50–53. Bibcode:2009Sci...324...50T. doi:10.1126/science.1161256.

- Sharp, Warren D.; Clague, David A. (2006). "50-Ma initiation of Hawaiian-Emperor bend records major change in Pacific Plate motion". *Science* **313** (5791): 1281–1284. Bibcode:2006Sci...313.1281S. doi:10.1126/science.1128489. PMID 16946069.

- Wilson, J. Tuzo (1963). "A possible origin of the Hawaiian Islands" (PDF). *Canadian Journal of Physics* **41**: 863–870. Bibcode:1963CaJPh..41..863W. doi:10.1139/p63-094.

- Ken Rubin, "The Formation of the Hawaiian Islands"

- USGS, "The long trail of the Hawaiian hotspot"

- National Geographic News: John Roach, "Hot Spot That Spawned Hawaii Was on the Move, Study Finds" : August 14, 2003

- Evolution of Hawaiian Volcanoes from the USGS.

- The Formation of the Hawaiian Islands with tables and diagrams illustrating the progressive age of the volcanoes.

- Hot Spots and Mantle Plumes

3.9   Jingū Seamount

Jingū Seamount, also called **Jingū Guyot**, is a guyot of the Hawaiian-Emperor seamount chain in the Pacific Ocean. It erupted 55 million years ago. The seamount is elongated in structure, running North-South, and has an oval shaped crater in the center, which is evidence of collapse when above sea level.[*][1][*][2]

The seamont was named in 1954 by Robert S. Dietz,[*][2] after Japanese Empress Jingū.[*][2]

3.9.1   See also

- List of volcanoes in the Hawaiian – Emperor seamount chain

3.9.2   References

[1] Jingu Seamount - John Search

[2] Calgue, David A.; Dalrymple, G. Brent; Greene, H. Gary; Wald, Donna; Kono, Masaru; Kroenke, Loren W. (1980). "40. Bathymetry of the Emperor Seamounts" (PDF). *Initial Reports of the Deep Sea Drilling Project* **55**. Washington, DC: US Government Printing Office. pp. 846–847. LCCN 74-603338. Retrieved April 25, 2012.

Coordinates: 38°45′N 171°15′E / 38.750°N 171.250°E

3.10   Kammu Seamount

Kanmu Seamount is a seamount lying within the Hawaiian-Emperor seamount chain in the Pacific Ocean. The last eruption of Kanmu Seamount is unknown.[*][1]

3.10.1   See also

- List of volcanoes in the Hawaiian – Emperor seamount chain

3.10.2   References

[1] Kanmu Seamount - John Search

Coordinates: 32°00′N 173°10′E / 32.000°N 173.167°E

3.11   Kimmei Seamount

Kimmei Seamount is a seamount of the Hawaiian-Emperor seamount chain in the northern Pacific Ocean. It last erupted about 40 million years ago.[*][1]

3.11.1   See also

- List of volcanoes in the Hawaiian – Emperor seamount chain

3.11.2   References

[1] Kimmei Seamount - John Search

Coordinates: 35°00′N 171°40′E / 35.000°N 171.667°E

3.12   Kodiak Seamount

Kodiak Seamount is the oldest seamount in the Kodiak-Bowie Seamount chain, with an estimated age of 24 million years.[*][1] It lies at the northernmost end of the chain and its flat-topped summit is strewn with fault lines. Like the rest of the Kodiak-Bowie seamounts, it was formed by the Bowie hotspot.

Kodiak Seamount will eventually be destroyed by subduction by the Aleutian Trench once it is carried into the trench by the ongoing plate motion, although this will not fully occur for several million more years if the current rate of motion is maintained. Because of Kodiak Seamount's approach into the Aleutian Trench, it is literally cracking up under the stress.[*][2] Although Kodiak is the oldest extant seamount in the Kodiak-Bowie chain, the adjacent lower slope contains transverse scars indicating earlier subduction of seamounts.

3.12.1   See also

- Bowie Seamount

3.12.2   References

[1] NOAA Ocean Explorer: Gulf of Alaska 2004 Retrieved on 2007-09-03

[2] Alaska Demo: Day 3, July 31, 1999 Retrieved on 2007-09-16

3.13   Kodiak–Bowie Seamount chain

The **Kodiak–Bowie Seamount chain**, also called the **Pratt–Welker Seamount chain**, is a seamount chain in southeastern Gulf of Alaska stretching from the Aleutian Trench in the north to Bowie Seamount, the youngest volcano in the chain, which lies 180 km (112 mi) west of the Queen Charlotte Islands, British Columbia, Canada.[*][1] The oldest volcano in the chain is the Kodiak Seamount. Although the Kodiak Seamount is the oldest extant seamount in the Kodiak-Bowie chain, the adjacent lower slope contains transverse scars indicating earlier subduction of seamounts.

The Kodiak–Bowie Seamount chain are mostly extinct volcanoes that formed above the Bowie hotspot. This is a 100- to-150-km-wide morphological swell presumably of thickened hotspot generated crust, although there are no seismic refraction data across the swell to define crustal thickness. The crest of one such peak, Patton Seamount originally formed off Washington state as a submerged volcano 33 million years ago. Over time, as the Pacific Plate moved steadily northwest, Patton Seamount was carried off the Bowie hotspot and into the Gulf of Alaska. New volcanoes were formed one after another over the hotspot, creating the Kodiak–Bowie Seamount chain.

Explorations of the Kodiak–Bowie Seamount chain have shown that despite the fact that most of the seamounts were created by the Bowie hotspot, all are unique in their size, shape, and volcanic features. The seamounts teem with deep-sea corals, sponges, and fish. Recent expeditions to these seamounts using manned submersibles and ROVs have discovered many marine species and have greatly expanded the knowledge of the range of deep sea corals in this region. For example, the Bowie Seamount is a biologically rich area with a dynamic and productive ecosystem. Because of this unique biological rich area, Bowie Seamount was declared a Pilot Marine Protected Area on December 8, 1998.*[2]

The Kodiak–Bowie seamount chain is at the northern triple junction between the Pacific, North American, and Juan de Fuca plates. Available age determinations on Kodiak and Giacomini Seamounts give an approximate average rate of movement along the chain of 6.5 cm (3 in) per year.

3.13.1   Volcanoes

Volcanoes in the chain include:

- Kodiak Seamount (24 m.y. old)*[3]

- Giacomini Seamount (21 m.y. old)*[3]

- Pratt Seamount (age unknown)*[3]

- Welker Seamount (15 m.y. old)*[3]

- Dickens Seamount (~4 m.y. old)*[3]

- Denson Seamount (18 .m.y. old)*[3]

- Durgin Seamount

- Brown Seamount

- Quinn Seamount

- Surveyor Seamount

- Peirce Seamount

- Hodgkins Seamount

- Bowie Seamount (>0.7 m.y. old)*[3]

- Tuzo Wilson Seamounts

3.13.2   References

[1] The Bowie Seamount Retrieved on 2007-09-02

[2] Bowie Seamount Marine Protected Area

[3] NOAA Ocean Explorer: Gulf of Alaska 2004 Retrieved on 2007-09-02

3.14 Koko Guyot

Koko Guyot (also sometimes known as *Kinmei*[2] and Koko Seamount) is a 48.1-million-year-old guyot,[3] a type of underwater volcano with a flat top, which lies near the southern end of the Emperor seamounts, about 200 km (124 mi) north of the "bend" in the volcanic Hawaiian-Emperor seamount chain.[5] Pillow lava has been sampled on the north west flank of Koko Seamount, and the oldest dated lava is 40 million years old.[4] Seismic studies indicate that it is built on a 9 km (6 mi) thick portion of the Pacific Plate.[6] The oldest rock from the north side of Koko Seamount is dated at 52.6 and the south side of Koko at 50.4 million years ago. To the southeast of the bend is Kimmei Seamount at 47.9 million years ago and southeast of it, Daikakuji at 46.7.[7]

3.14.1 Geology and characteristics

The seamount was named for the 58th emperor of Japan, Emperor Koko (A.D. 885-887) by geologist Thomas Davies and his colleagues in 1972, based on the results from a bathymetric expedition and contents of two dredge huals, led by Thomas Washington and undertaken with the ship *Aries-7*.[1][5] The seamount is elongate in shape, aligned northwest-southeast (the same direction as the chain), and has a gentle slope and a large, flat top. Koko Seamount also has a lot of small reefal bodies on its slopes.[1] It rises from the abyssal floor about 5,000 m (16,000 ft) in height.

A prominent south-trending ridge extends about 50 km (31 mi) from the summit area in the direction of Kimmei Seamount, to the southeast.[5] The base of the guyot is similar to a "pedestal," and is composed of consolidated lavas and extinct volcanic centers of the volcano's formally active history; it is similar to structure to the pedestal found at the base of most of the other, usually larger Emperor seamounts. However, a thick carbonate cap, similar to the one covering Detroit Seamount, makes it difficult to find the exact eruptive centers.[5] The volcano is clearly isolated, even in comparison to other seamounts in the spread-out Emperor chain, with Ojin Seamount about 200 km (124 mi) to the northwest and Kimmei Seamount 100 km (62 mi) to the southeast.[5] The seamount is located just 2.3 degrees north of the bend.[1]

Much of what we know about Koko comes from early dredgings and the Ocean Drilling Program's core samples, collected as part of Leg 197, at Site 1206, which aimed to supply information on the relatively obscure Emperor seamounts and study their relation to the Hawaiian chain.[1][8] Site 1206 was the last and southernmost drilling site during Leg 197, and was located on the southeastern side of the lower summit terrace of Koko Seamount.[1] A seismic survey of the region was utilized to locate a suitable place for the drill site, initially targeted near Site 308, drilled in 1973 during Leg 32. Weather conditions during the drilling had prevented it from reaching 68.5 m (225 ft) in depth, the approximate depth of the sediment cover in the region.[1] Due to a shortage of time, priority was placed on finding a region with a thin sedimentary cover. The site eventually chosen was located at a water depth of 1,545 m (5,069 ft), 6.2 km (4 mi) south of Site 308, at coordinates 34°55.55′N 172°8.75′E / 34.92583°N 172.14583°E. The sediment cover at this site was less than half that at the 1973 drill site, and rock was hit at a subsurface depth of 57 m (187 ft). Drilling continued to 278 m (912 ft) into the slopes.[1]

The top 57 m (187 ft) of sediment included fossil-rich calcarenite and calcium-rich mudstone and siltstone, indicating a shallow-water setting at the time of deposition.[1] The lower part of the core sample recovered a 15 cm (6 in) to 20 cm (8 in) section of shell-bearing mudstone containing many microfossils typical of the early to middle Eocene (43.5-49.7 Ma). This age range fits well with a radiometric analysis (48.1 Ma) reported for a dredged rock from Koko Seamount from the 1973 expedition. Although shell fragments had been recovered from the sediment cover in 1973, none of these deposits contained microfossils.[1]

Lava flows dominate the lithology of the main body, with a small proportion of calcarenite. Many lavas were pahoehoe flows laced with a'a, evidence of subaerial eruptions.[1] There was a large amount of variation in the density, structure, porosity, and grain size of the recovered volcanic rock, varying widely with depth. The bulk of the volcanic rock is basalt of aphyric to olivine-phyric lava, and tholeiitic or alkalic in composition. The basaltic lavas from Koko Seamount resemble those drilled during Leg 55, at Suiko Seamount.[1]

Studies suggested that the magnetic arrangement of the rock, used to determine its latitude at formation (magnets align to the North pole; also, the drift and position of the Hawaii hotspot at various times is important to hotspot studies), were relatively stable. 14 magnetic groupings were found on the seamount, yielding a mean latitude of 38.5 degrees south of the seamount's present location (the percent of error is +8.4°/−10.9°). That would put the seamount at 21.7° N in latitude during its early history, before the Pacific Plate moved it to its current position relative to Earth.[*][1]

3.14.2 Ancient ecology

Dredged carbonate samples from the top of the seamount contained porites and several other corals, covered by coralline algae at shallow to medium depth. Also present were Amphistegina, red algae (mainly Lithothamnion and Sporolithon), lepidocyclines, bryozoans, and coralline at deeper depths. The recorded lepidocyclinids indicate an Early Miocene age for the drowned carbonate platforms found on the seamount, at about 500 m (1,640 ft).[*][9]

3.14.3 See also

- Hawaii hotspot

- Hawaiian-Emperor seamount chain

3.14.4 See also

- List of volcanoes in the Hawaiian – Emperor seamount chain

3.14.5 References

[1] "SITE 1206". *Ocean Drilling Program Database-Results of Site 1206*. Ocean Drilling Program. Retrieved 2009-04-09.

[2] "Seamount Catalog". *Seamounts database*. EarthRef, a National Science Foundation project. Retrieved 2009-04-09.

[3] Dyar, Darby. "HOTSPOTS AND PLATE MOTION". Retrieved 2009-04-04.

[4] Seach, John. "Koko Seamount, NW Flank - John Seach". *Volcanic database*. Volcano Live.com. Retrieved 2009-04-09.

[5] "6. Site 12061 BACKGROUND AND SCIENTIFIC OBJECTIVES". *Drilling Site Reccomendation Submission for Koko*. Ocean Drilling Program. Retrieved 2009-04-09.

[6] K. FURUKAWA, J. F. GETTRUST, L. W. KROENKE and J. F. Campbell (1980). "Crust and upper mantle structure along the flank of Koko Seamount". *Scientific Paper-Abstract*. Hawaii Institute of Geophysics University of Hawaii, Honolulu, Hawaii 96822. Retrieved 2009-04-09.

[7] TenBruggencate, Jan (2006). "Hawaiian geology gets update". *Honolulu Advertiser web article*. Honolulu Advertiser. Retrieved 2009-04-09.

[8] "DRILLING STRATEGY". *Ocean Drilling Program - Leg 197 Proposal*. Ocean Drilling Program. Retrieved 2009-04-09.

[9] David A. Clauge , Juan C. Braga, Jody M. Webster , Davide Bassi , Willen Renema. "Lower Miocene submerged reefs on the Koko Seamount". *Essay Abstract*. Retrieved 2009-04-09.

3.15 Lord Howe seamount chain

The **Lord Howe seamount chain** is the seamount chain that includes Lord Howe Island. [*][1] It is one of the two parallel seamount chains alongside the east coast of Australia; the Lord Howe and Tasmantid seamount chains both run north-south through parts of the Coral sea and Tasman sea.[*][1] These chains have longitudes of approximately 159°E and 156°E respectively.[*][1]

The Lord Howe seamount chain is on the western slope of Lord Howe Rise, a deep-sea elevated plateau which is a submerged part of Zealandia.[*][1] The Tasmantid and Lord Howe seamount chains are both broadly within the Tasman basin (the abyssal plain between Lord Howe Rise and the Australian continental shelf), and lie on opposite sides of Dampier Ridge (a submerged continental fragment).[*][2][*][1]

The Lord Howe seamount chain extends from the Chesterfield group (20°S) to Flinder's seamount (34.7°S).*[3] It includes Nova bank, Argo and Kelso seamounts, Capel and Gifford guyots, Middleton and Elizabeth reefs, Lord Howe Island and Balls Pyramid.*[1]

The Lord Howe and Tasmantid chains each resulted from the Indo-Australian Plate moving northward over a stationary hotspot; the hotspot for the Lord Howe chain is expected to presently be beneath Flinder's seamount.*[4] On the Australian mainland, a third north-south sequence of extinct volcanoes (which includes the Glasshouse mountains) is likely to have the same origin.*[4]

The chain formed during the Miocene. It features many coral-capped guyots.

3.15.1   References

[1]   Willem J. M. van der Linden, *Morphology of the Tasman sea floor*. New Zealand Journal of Geology and Geophysics. Vol.13 (1970) 282-291.

[2]   McDougall et al, *Dampier Ridge, Tasman Sea, as a stranded continental fragment*. Australian Journal of Earth Sciences 41 (1994). 395-406.

[3]   Przeslawski et al. *Biogeography of the Lord Howe Rise region, Tasman Sea*. Deep-Sea Research II 58 (2011) 959–969.

[4]   W. J. Morgan and J. P. Morgan. *Plate velocities in hotspot reference frame: electronic supplement.*

3.16   Louisville seamount chain

The **Louisville seamount chain** is an underwater chain of over 70 seamounts in the Southwest Pacific Ocean. As one of the longest seamount chains on Earth it stretches some 4,300 kilometres*[1] from the Pacific-Antarctic Ridge north west to the Tonga-Kermadec Trench, where it subducts under the Indo-Australian Plate as part of the Pacific Plate. The movement of the Pacific Plate over the Louisville hotspot formed the chain.

Depth-sounding data first revealed the existence of the seamount chain in 1972.*[2]

3.16.1   See also

- Hotspot (geology)

- Osbourn Seamount the oldest seamount in the Louisville chain

3.16.2   References

[1]   Vanderkluysen, L.; Mahoney, J. J.; Koppers, A. A.; and Lonsdale, P. F. (2007). Geochemical Evolution of the Louisville Seamount Chain, American Geophysical Union, Fall Meeting 2007, abstract #V42B-06.

[2]   Sandwell, David T.; Walter H.F. Smith (1997). "Exploring the ocean basins with satellite altimeter data". *Satellite Geodesy*. La Jolla: Scripps Institution of Oceanography. Retrieved 2010-01-19. The Louisville seamount chain was first detected in 1972 using depth soundings collected along random ship crossings of the South Pacific. Six years later the full extent of this chain was revealed by a radar altimeter aboard the Seasat (NASA) spacecraft.

3.16.3   External links

- Expedition 330 - Louisville Seamount Trail, Integrated Ocean Drilling Program, 13 December 2010 to 11 February 2011

- The Louisville Ridge – Tonga Trench collision: Implications for subduction zone dynamics, RV Sonne Research Expedition SO215 Cruise Report, 25 April 2011 to 11 June 2011

3.17   Meiji Seamount

Meiji Seamount, named after Emperor Meiji, the 122nd Emperor of Japan, is the oldest seamount in the Hawaiian-Emperor seamount chain, with an estimated age of 82 million years. It lies at the northernmost end of the chain, and is perched at the outer slope of the Kuril-Kamchatka Trench. Like the rest of the Emperor seamounts, it was formed by the Hawaii hotspot volcanism, grew to become an island, and has since subsided to below sea level, all while being carried first north and now northwest by the motion of the Pacific Plate. Meiji Seamount is thus an example of a particular type of seamount known as a guyot, and some publications refer to it as **Meiji Guyot**.

Meiji Seamount will eventually be destroyed by subduction by the Aleutian Trench once it is carried into the trench by the ongoing plate motion, although this will not fully occur for several million more years if the current rate of motion is maintained. Although Meiji is the oldest extant seamount in the Hawaii-Emperor chain, the question of whether there were older seamounts in the chain which have already been subducted into the trench remains open, and is the subject of ongoing scientific research.

The Deep Sea Drilling Project (DSDP) Leg 19, Hole 192A, recovered 13 m (43 ft) of pillow lava from near the summit of Meiji.*[1] The lavas were initially classified as alkali basalts on the basis of their mineralogy, but subsequent microprobe analyses of glass and pyroxene suggested that they are tholeiitic in origin. At least five flows were found.*[2]

3.17.1   See also

- Hawaiian-Emperor seamount chain

- Evolution of Hawaiian volcanoes

- Hotspot (geology)

- Detroit Seamount

3.17.2   References

Notes

[1] (Scholl et al. 1973)

[2] (Regelous et al. 2003)

Bibliography

- "Volcano World: The Hawaiian - Emperor Volcanic Chain" . Retrieved 2007-03-25.

- Scholl, David W.; Creager, Joe S.; Boyce, Robert E.; Echols, Ronald J.; Fullam, Timothy J.; Grow, John A.; Koizumi, Itaru; Lee, Homa J.; Ling, Hsin Yi; Supko, Peter R.; Worsley, Thomas R. (1973). "Site 192: Meiji Seamount" (PDF). *Deep Sea Drilling Project Initial Reports* **19**: 463–533. doi:10.2973/dsdp.proc.19.111.1973.

- Dalrymple, G. Brent; Lanphere, Marvin A.; Natland, James H. (1980). "K-Ar Minimum Age for Meiji Guyot, Emperor Seamount Chain" (PDF). *Deep Sea Drilling Project Initial Reports* **55**: 677–683. doi:10.2973/dsdp.proc.55.129.1980.

- Scholl, D. W.; Rea, D. K. (2002). "Estimating the Age of the Hawaiian Hotspot" . *American Geophysical Union* **61**: 05. Bibcode:2002AGUFM.T61C..05S.

- Regelous, M.; Hofmann, A. W.; Abouchami, W.; Galer, S. J. G. (2003). "Geochemistry of Lavas from the Emperor Seamounts, and the Geochemical Evolution of Hawaiian Magmatism from 85 to 42 Ma" . *Journal of Petrology* (Max-Planck Institute for Chemie, Abteilung Geochemie, Post 3060, 55020 Mainz, Germany: Oxford University Press) **44** (1): 113–140. doi:10.1093/petrology/44.1.113.

- Norton, Ian O. (2006). "Speculations on tectonic origin of the Hawaii hotspot". Mantleplumes.org. Retrieved 2014-08-08.

- Shapiro, M. N.; Soloviev, A. V.; Ledneva, G. V. (2006). "Did Emperor seamounts subduct?". Mantleplumes.org. Retrieved 2014-08-08.

3.18 Nintoku Seamount

Nintoku Seamount or **Nintoku Guyot** is a seamount (underwater volcano) and guyot (flat top) in the Hawaiian-Emperor seamount chain. It is a large, irregularly shaped volcano that last erupted 66 million years ago. Three lava flows have been sampled at Nintoku Seamount; the flows are almost all alkalic (subaerial) lava.[4] It is 56.2 million years old.[3]

Nintoku is positioned a roughly 41 degrees north latitude, approximately two-thirds the way southward along the north-northeast-south-southeast Emperor seamounts extending from Meiji Seamount (about 53°N) in the north to Kammu Seamount (about 32°N) at the chain's southern terminus. Nintoku Seamount was named after the 16th emperor of Japan, Emperor Nintoku, by geologist Robert Dietz in 1954.[4]

The seamount occupies a central position in the Emperor Seamount chain and is thus an important point in the paleolatitude history of the Hawaiian hotspot, instrumental to proving the scientific hunch that the Hawaii hotspot was a mobile entity.[1] The structure of the seamount is elongate, aligned north-northwest along the Emperor trend, with two prominent ridges trending southwest and south-southwest as far as 100 km (62 mi) from the main crater. Nintoku Seamount is a plexus of coalesced volcanoes, much like many of the larger seamounts in this chain. The Nintoku system is, however, clearly isolated from Yomei Seamount, about 100 km (62 mi) to the north, and Jingu Seamount, about 200 km (124 mi) to south, by abyssal depths.[1]

3.18.1 Geology and characteristics

In seismic profile, the main body of the seamount rises steeply over 5,000 m (16,404 ft) in predominantly unsedimented volcanic slope to the thinly sedimented (10 m (33 ft) to 200 m (660 ft)), from an Emperor point of view, gently domed summit region between 1,200 m (3,937 ft) to 1,400 m (4,593 ft) high peak profile, which covers about 3400 square kilometers of area.[1]

From analysis of seismic reflection survey data and core material recovered by drillings at Site 432, the shipboard party of Deep Sea Drilling Project (DSDP) Leg 55 proposed that Nintoku Seamount was in an intermediate atoll stage (no lagoon but fringing reefs and banks and extensive carbonate bank interiors) before subsidence removed the island below the wave base. It was further thought that a few small remnant volcanic peaks and domes still pierce the sedimentary deposits.[1]

Nintoku Seamount apparently remained at or above sea level long enough to be almost completely devastated by subaerial erosion and wave action. Reefs were not indicated in the seismic studies, but fragmented pieces of coral were recovered and documented, showing a shallow-water sediment-rich condition. The rock records indicate deposition in waters cooler than the present tropical condition. Shallow-water sedimentary deposition ceased in Paleogene times.[1]

The seamount was first drilled by Site 432, located on the northwestern edge of the summit region of Nintoku Seamount, in a gently sloping area mapped as terrace deposits. Although the sediment cover was estimated, based on other seamount covers, to be 80 meters (262 ft) thick, bedrock was hit after only 42 m (138 ft). Poorly recovered and preserved sedimentary deposits indicated a shallow-reef bed typical of terraced flanking reefs and banks, as well as volcanic sand. Drilling at Site 432 penetrated 32 m (105 ft) of volcanic rock (74 m (243 ft) total) before terminating because of hole caving and damage to the drill assembly.[1]

The site was drilled as part of Leg 197 by the Ocean Drilling Program, at site number 1205.[4] A short bathymetric acoustic survey was conducted to find the best site for the location and structure of a core sampling. The locale chosen was about 100 m (328 ft) southwest of Site 432, the location of a previous drilling by the ODP.

Site 1205 (41°20.00′N 170°22.70′E / 41.33333°N 170.37833°E)[1] was located in 1,310 m (4,298 ft) deep water, where previous drilling had reached volcanic rock beneath Paleocene deposits. It was elected to return to the site for a number of reasons. First, drilling at the nearby site 432 had hit reasonably unaltered and unchanged basalt with good remnant magnetic properties, key to finding the latitude of origin; but insufficient sampling caused a lack of data, and determining the age accurately was not possible. Hence, deeper drilling was promised to achieve that goal, providing a time-averaged (as the seamount is in the center of the chain) movement ratio.[*][1] Second, a survey of the region showed a rock structure suitable for deep drilling, and nearby sites met low levels of sedimentary cover. Thirdly, the composition of previously drilled volcanic rock seemed to match the volcano's "average" type, erupted during the post-shield stage of it life. This helped another project goal, to recover a suitable and datable chunk of tholeiitic lava, which appeared to be rare on the seamount.[*][1]

The hole was drilled in what appeared to be a large, broad sedimentary cover (estimated 70 m (230 ft) in thickness) covering a swath of the ancient volcano's main slope.[*][4] The coring encountered volcanic rock at 42 m (138 ft) below the sea floor, and continued to a final depth of 326 m (1,070 ft) below the sea floor. The sedimentary cover, an element commonly found on many of the Emperor seamounts, was found to be largely a stack interlaced lava flows (about 95%). The drilling penetrated 283 m (928 ft) into the seamount's volcanic rock, and recovered at least 25 different hardened lava flows.[*][4]

It was established that Nintoku Seamount's sedimentary cap consists of sandstone and siltstone containing well-rounded to subrounded basalt blocks, volcanic ash, fossil fragments of mollusks, benthic foraminifers, bryozoans, and coralline red algae. These observations indicate a shallow-water, high-rate depositional setting. Little variation was found in the density, grain size, or porosity of the volcanic rock, and it was stable in composition, except for the volcanic-sedimentary cover. It is believed that this is the underlying cause of aucousically recorded layering of the upper 200 m (656 ft)–230 m (755 ft) of rock, after which the effect of soil layering fades away.[*][4]

The age of the youngest volcanic rocks was constrained by nano-fossils in the sediment to be older than 53.6–54.7 million years, an age that is just younger than the radiometric age of 56.2 ± 0.6 million obtained for the basalt drilled from nearby Hole 432A.[*][4] The thickness and vesicularity of the lava flows, as well as the presence of oxidized flow tops and soil horizons and a lack of pillow lava, indicate that all of the obtained samples erupted subaerially. The volcanic rock ranges from aphyric to highly plagioclase and olivine-phyric basalt. At 230 to 255 meters below ground, two flows of tholeiitic basalt were found interlaced with the alkalic basalt flows. Above these flows the degree of alkalinity skyrockets.[*][4] There is evidence suggesting that the eruption rates must have been lower at the period during which the two flows were deposited, which is consistent with the model of Hawaiian volcanic growth, which increases in activity slowly over time before ceasing altogether. Internal lava flows have also created course-grained rocks.[*][4] Lavas from Nintoku Seamount have similar composition to lava erupted during the post-shield stage of Hawaiian volcanoes such as Mauna Kea. Slight differences in trace element composition between lavas from Nintoku Seamount and active Hawaiian volcanoes probably result from differences in source composition or variations in the degree of melting.[*][4]

All of the recovered lava flows had been altered very little by erosion or other lava flows, except for thin flow tops. Sparse veining indicates that there is only small-scale fluid circulation within the rocks, in contrast with some of the data collected from Detroit Seamount.[*][4]

Rock magnetic data obtained suggest that the lava flows from Site 1205 carry a remnant magnetization suitable for scientific analysis. Although some of the rocks needed a more complex and thorough analysis, most samples yielded data suitable to make a preliminary determination of magnetic inclinations. Twenty-two independent magnetic group were identified, yielding a mean inclination of $-45.7°$ (+10.5°/−6.3°). The mean inclination suggests a latitude of formation on an early Eocene Nintoku Seamount at 27.1° (+5.5°/−7.7°).[*][4] This value, together with paleo-latitudes from analyses of rocks at Site 433 (1980), Site 884 (1997), and Sites 1203 and 1204 (Leg 197; Detroit Seamount), form a consistent data set indicating southward motion of the Hawaiian hotspot from Late Cretaceous to early Tertiary time, a hunch that many scientists had harbored for a long time.[*][4]

3.18.2   See also

- List of volcanoes in the Hawaiian – Emperor seamount chain

3.18.3   References

[1] "Site 1205 Background and Scientific Objectives". *Ocean Drilling Program database entry*. Ocean Drilling Program. Retrieved 2009-04-10.

[2] "Seamount Catalog". *Seamounts database.* EarthRef, a National Science Foundation project. Retrieved 2009-04-10.

[3] "Age of Hawaiian-Emperor Volcanoes as a Function of Distance from Kilauea". *Graphic Representation of Ages.* Enduring Resources for Earth Science Education (ERESE). Retrieved 2009-04-04.

[4] "SITE 1205 Principal Results". *Ocean Drilling Program Entry.* Ocean Drilling Program. Retrieved 2009-04-10.

3.19   Ojin Seamount

Ōjin Seamount, also called **Ōjin Guyot**, named after Emperor Ōjin, 15th Emperor of Japan, is a guyot of the Hawaiian-Emperor seamount chain in the Pacific Ocean.[1] It erupted 55 million years ago.[2]

3.19.1   See also

- List of volcanoes in the Hawaiian – Emperor seamount chain

3.19.2   References

[1] (Clague et al. 1980, p. 848)

[2] (Seach 2014)

Bibliography

- Seach, John. "Ōjin Seamount - John Seach". *Volcanic database.* Volcano Live.com. Retrieved 2014-00-09. Check date values in: |access-date= (help)

- Jackson, Everett D.; Koizumi, Itaru; Kirkpatrick, R. James; Avdeiko, Gennady; Clague, David A.; Dalrymple, G. Brent; Karpoff, Anne-Marie; McKenzie, Judith; Butt, Arif; Ling, Hsin Yi; Takayama, Toshiaki; Green, H. Gary; Morgan, Jason; Kono, Masaru (1980). "Site 430: Ōjin Seamount" (PDF). *Deep Sea Drilling Project Initial Reports* **55**: 45–76. doi:10.2973/dsdp.proc.55.103.1980.

- Clague, David A.; Dalrymple, G. Brent; Greene, H. Gary; Wald, Donna; Kono, Masaru; Kroenke, Loren W. (1980). "Bathymetry of the Emperor Seamounts" (PDF). *Deep Sea Drilling Project Initial Reports* **55**: 845–849. doi:10.2973/dsdp.proc.55.140.1980.

- Watts, A. B.; Ribe, N. M. (December 10, 1984). "On Geoid Heights and Flexure of the Lithosphere at Seamounts" (PDF). *Journal of Geophysical Research* **89** (B13): 11152–11170. doi:10.1029/jb089ib13p11152.

Coordinates: 38°15′N 171°00′E / 38.250°N 171.000°E

3.20   Osbourn Seamount

Osbourn Seamount is the oldest seamount in the Louisville seamount chain, with an estimated age of 77.3 million years.[1] It lies at the westernmost end of the chain and its flat-topped summit is strewn with fault lines. Like the rest of the Louisville seamount chain, it was formed by the Louisville hotspot.

Osbourn Seamount will eventually be destroyed by subduction by the Tonga and Kermadec trenches once it is carried into the trenches by the ongoing plate motion, although this will not fully occur for several million more years if the current rate of motion is maintained.

3.20.1   References

[1] Plates, Plumes, And Paradigms

3.21   Rodriguez Seamount

Rodriguez Seamount is a seamount and guyot located about 150 km (93 mi) off the coast of Central California. It is structurally similar to the nearby Guide, Pioneer, Gumdrop, and Davidson seamounts, all located roughly between 37.5° and 34.0° degrees of latitude. This group of seamounts is morphologically unique, and the mounts are very similar to one another. The seamount structures run parallel to an ancient spreading center which has since been replaced in its role by the San Andreas Fault system.*[3]

3.21.1   Geology

Magnetic anomalies at Rodriguez indicate that it is located on a 19-million-year-old crustal surface. Rocks recovered from Rodriguez Seamount are largely composed of alkaline basalt and Hawaiite. Ar-Ar dating techniques indicate that the volcano is between 10 and 12 million years of age.*[1]

Rodriguez Seamount rises about 1,675 m (5,495 ft) above the surrounding seafloor, to a minimum depth of 650 m (2,133 ft). Its calculated volume is greater than 205 km^3 (49 cu mi); however this is likely an understatement because the survey did not include its lowermost slope.*[1]

The slope is composed mostly of layered volcanic rock, mostly coarse sandstone, with a few scattered large lava boulders. They were likely formed from fragments of volcanic glass formed in the steam explosions of lava touching down against water, similar to the process that is happening today on Kilauea. This would have built a black sand beach over time; however, following millions of years of alterations, most of the sand has since been converted into clay. Small hills, constructed of jagged lava flows, are thought to have resulted from subaerial 'a'a flows.*[4]

The northeast-trending ridges, which are common to the group that Rodriquez Seamount is in, are less distinct on Rodriguez than on the other seamounts in the group. In addition, the seamount is propagated by several large volcanic cones, the largest of which is 700 m (2,297 ft) tall and 2.2 km (1.4 mi) at the base, with a volume of about 2.6 km^3 (1 cu mi). The lower flanks of the volcano have slumped and are covered in a thick layer of sediment, particularly the southwest flank. Another slump area to the west flank has blocky debris on it, suggesting that the base of the volcano has also started collapsing into itself.*[1]

Rodriguez Seamount once extended above sea level, resulting in a flat, sediment-covered summit that is coated with beach sands of ancient origin. This flat top earns it the distinction of being a guyot. These sands have been colonized by, among others, sea cucumbers.*[5]

An expedition in 2003 included Rodriguez Seamount as one of its destinations. Observations made during the expedition confirmed theories that the seamount had once been above sea level, and has since subsided about 750 m (2,461 ft) from its former height. A sandstone structure discovered at 650 m (2,133 ft) depth seems to hint at a former sand beach and shoreline.*[4]

3.21.2   References

[1] "Mapping Program: Rodriguez Seamount". MBARI. Feb 6, 2009. Retrieved 8 December 2009.

[2] "Seamount Catalog". *Seamounts database*. EarthRef, a National Science Foundation project. Retrieved 2009-04-09.

[3] "Geology of Davidson Seamount". NOAA, Office of Ocean Exploration and Research. February 3, 2006. Retrieved December 2, 2009.

[4] Clauge, David (October 13, 2003). "Cruise In The Classroom: Seamounts 2003 October 11- October 17, 2003". MBARI. Retrieved 8 December 2009.

[5] "Seamounts may serve as refuges for deep-sea animals that struggle to survive elsewhere". PhysOrg. February 11, 2009. Retrieved December 7, 2009.

3.22 Suiko Seamount

Suiko Seamount, also called **Suiko Guyot**, is a guyot of the Hawaiian-Emperor seamount chain in the Pacific Ocean.

The last eruption from Suiko Seamount occurred 60 million years ago, during the Paleogene Period of the Cenozoic Era.*[1]*[2]

3.22.1 References

[1] Suiko Seamount,North - John Search

[2] Suiko Seamount, Central Region - John Search

Bibliography

- Jackson, Everett D.; Koizumi, Itaru; Kirkpatrick, R. James; Avdeiko, Gennady; Clague, David; Dalrymple, G. Brent; Karpoff, Anne-Marie; McKenzie, Judith; Butt, Arif; Ling, Hsin Yi; Takayama, Toshiaki; Green, H. Gary; Morgan, Jason; Kono, Masaru (1980). "Site 433: Suiko Seamount" (PDF). *Deep Sea Drilling Project Initial Reports* **55**: 127–282. doi:10.2973/dsdp.proc.55.106.1980.

Coordinates: 44°35′N 170°20′E / 44.583°N 170.333°E

3.23 Yomei Seamount

Yomei Seamount is a seamount of the Hawaiian-Emperor seamount chain in the northern Pacific Ocean.

Its eruption ages are unknown, but the seamonts on either side are in the 56.2 to 59.6 million range during the Paleogene Period.*[1]

3.23.1 See also

- List of volcanoes in the Hawaiian – Emperor seamount chain

3.23.2 References

[1] Yomei Seamount - John Search

Bibliography

- Jackson, Everett D.; Koizumi, Itaru; Kirkpatrick, R. James; Avdeiko, Gennady; Clague, David; Dalrymple, G. Brent; Karpoff, Anne-Marie; McKenzie, Judith; Butt, Arif; Ling, Hsin Yi; Takayama, Toshiaki; Green, H. Gary; Morgan, Jason; Kono, Masaru (1980). "Site 431: Yōmei Seamount" (PDF). *Deep Sea Drilling Project Initial Reports* **55**: 77–94. doi:10.2973/dsdp.proc.55.104.1980.

Coordinates: 42°20′N 170°07′E / 42.333°N 170.117°E

3.24   Yuryaku Seamount

Yuryaku Seamount (also called *Yuryaku Guyot*) is a seamount (underwater volcano) and guyot (flat-topped) located northwest of Hawaii. It is located a little southwest of the V-shaped bend separating the Emperor Seamounts from the older Hawaiian islands, all of the Hawaiian-Emperor seamount chain in the North Pacific Ocean.

3.24.1   Geology

Alkalic basalt dredged from Yuryaku Seamount is similar to the alkalic basalt that caps the volcanoes in the Hawaiian Islands. Analyses gave a mean age of 42.3 ± 1.6 m.y. for Yuryaku Seamount.[*][2] The data collected helped show that the age of the Hawaiian-Emperor bend is about 41 to 43 m.y. Alkalic basalt have been sampled at Yuryaku Seamount.

The last eruptions of Yuryaku Seamount was 43 million years ago, during the Eocene epoch of the Paleogene Period.[*][3]

3.24.2   See also

- List of volcanoes in the Hawaiian – Emperor seamount chain

3.24.3   References

[1] "Seamount Catalog". *Seamounts database.* EarthRef, a National Science Foundation project. Retrieved 2009-04-10.

[2] Clague, D.A., Dalrymple, G.B. and Moberly, R. (1975). "Petrography and K-Ar ages of dredged volcanic rocks from the western Hawaiian Ridge and the southern Emperor Seamount Chain". *Society of America, Bulletin 86(7).* The Geological Society of America. pp. 991–998. Retrieved 2009-04-11.

[3] Seach, John. "Yuryaku Seamount - John Seach". *Volcanic database.* volcanolive.com. Retrieved 2009-04-11.

Chapter 4

Seamounts of the Pacific Ocean

4.1   2012 Kermadec Islands eruption

The 2012 Kermadec Islands eruption was a major undersea volcanic eruption that was produced by the previously little-known Havre Seamount near the L'Esperance and L'Havre Rocks[*][1] in the Kermadec Islands of New Zealand.[*][2][*][3] The large volume of low density pumice produced by the eruption accumulated as a large area of floating pumice, a *pumice raft,* that was variously estimated to be between 7,500 and 10,000 square miles (19,000 and 26,000 km^2).[*][4][*][5]

4.1.1   Eruption

The eruption of the Havre Seamount was not initially noticed by scientists, and volcanologists were not even aware that the Havre Seamount was an active submarine volcano.[*][2] After the pumice raft was detected, researchers retrospectively examined satellite imagery and past seismic activity in an attempt to pinpoint the time and location of the eruption that produced the pumice raft.[*][2] Seismologists discovered a cluster of earthquakes (ranging in magnitude between 3.0 and 4.8) that occurred 18–19 July 2012.[*][2][*][6] These earthquakes were consistent with magma rising into a magma chamber prior to eruption.[*][2] Analysis of satellite imagery showed an ash plume appear on 18 July 2012 and conclude several days later.[*][2][*][6] Although some volcanologists initially believed that the eruption might have occurred at the Monowai Seamount, this possibility was later ruled out.[*][6] It is also believed that the eruption was unrelated to the 2012 Te Māri eruption at New Zealand's Mount Tongariro.[*][7]

4.1.2   Pumice raft

Scientists were not aware that any eruption had occurred until a huge pumice raft was sighted and photographed at 14:40 NZST 0n 31 July 2012 by Maggie de Grauw while on a commercial flight from Faleolo, Samoa to Auckland, New Zealand. She emailed her pictures to Dr Scott Bryan, Senior Research Fellow at Queensland University of Technology. After discussion it was ascertained that the raft was around 1,000km north of Auckland. Dr Bryan then contacted Olivier Hyvernaud from the Laboratoire de Géophysique, Tahiti, who confirmed the location from Terra/MODIS imagery from NASA. The raft was subsequently sighted by members of the New Zealand Defence Force on 9 August 2012[*][8]—several weeks after the eruption had occurred.[*][2] It was spotted by an Orion aircraft and then approached and sampled by the strategic sealift ship HMNZS *Canterbury*.[*][6] The pumice raft measured approximately 300 miles (480 km) in length and more than 30 miles (48 km) in width, making the floating island larger in surface area than Israel.[*][7][*][8] An officer in the Royal Australian Navy was quoted as saying that it was "the weirdest thing [he had] seen in 18 years at sea".[*][7][*][8]

4.1.3   References

[1] Science Direct map of Kermadec Islands and Seamounts

[2] Klemetti, Erik (13 August 2012). "Havre Seamount: The Source of Kermadec Island Pumice Raft?". *Wired*. Retrieved 13 August 2012.

[3] Bryner, Jeanna (14 August 2012). "'Raft' in Pacific Found, NASA Says". *Live Science*. Retrieved 14 August 2012.

[4] Memmott, Mark (10 August 2012). "7,500 Square Miles Of Pumice Floating In Pacific Is 'Weirdest Thing I've Seen'". *NPR*. Retrieved 13 August 2012.

[5] AP (10 August 2012). "Massive Rock Raft Found Floating off New Zealand". *ABCNews.com*. Retrieved 13 August 2012.

[6] Cooke, Michelle (11 August 2012). "Scientists rock theory on pumice raft". *Stuff.co.nz*. Retrieved 13 August 2012.

[7] Gannon, Megan (10 August 2012). "Huge pumice rock 'island' seen floating in South Pacific: New Zealand's Royal Navy spots raft larger than Israel —but no one knows where it's from". *NBCNews.com*. Retrieved 13 August 2012.

[8] Gannon, Megan (13 August 2012). "Mystery Rock Shelf Floating in Pacific: Floating pumice rocks are covering an area larger than Israel in the South Pacific.". *Discovery News*. Retrieved 13 August 2012.

4.2   Adams Seamount

Adams Seamount is a submarine volcano above the Pitcairn hotspot in the central Pacific Ocean about 90 kilometres (56 mi) southwest of Pitcairn Island.

It is a massive seamount rising 3,500 metres (11,483 ft) from the sea floor to about 59 metres (194 ft) below the surface of the ocean.

4.2.1   References

• "Adams Seamount". *Global Volcanism Program*. Smithsonian Institution.

4.3   Bollons Seamount

Bollons Seamount is a seamount (underwater volcano) just east of the international date line, a few hundred miles off the coast of New Zealand.*[1] *[2]

The seamount was involved in a 2002 survey and collection project defined to find the end scope of the Australian Plate.*[3] The Bollons Seamount has been shown to be a site of extensive Cretaceous-era rifting in the area towards the southern Chatham Rise between 83.7 and 78.5 MYA. Magnetic anomalies from the seamount indicate that it was the site of highly irregular activity, with differences in the rifting there being up to 100 km (62 mi). A 50 km (31 mi) gap near the seamount, known as the Ballons gap, is interpreted as being due to excess volcanism is the seafloor-spreading process. A ridge just south of the seamount, the Antipodes Fracture Zone, is interpreted as having been built by a combination of compression and volcanic activity associated with the triple-junction Bellingshausen-Marie plate boundary nearby.*[4]

4.3.1   References

[1] "Bollons Seamount". *Seamount database*. Earthref, a National Science Foundation project.

[2] "Google Maps". Google. Retrieved 22 July 2010.

[3] "Hydrographic Report Number crunching time for Continental Shelf Project". 2002. Retrieved 22 July 2010.

[4] Davy, B. (2006), Bollons Seamount and early New Zealand–Antarctic seafloor spreading, Geochem. Geophys. Geosyst., 7, Q06021, doi:10.1029/2005GC001191.

4.4 Browns Mountain

For the mountain in South Carolina, USA, see Brown's Mountain.

Browns Mountain, also sometimes spelt **Brown's Mountain**, is a small submarine mountain – a seamount – lying in the south-western Pacific Ocean 38 km off the coast of New South Wales, Australia, about 50 km east of the city of Sydney. The waters around the seamount are about 600 m in depth, while the mountain itself rises some 133 m above the sea floor. It is a popular site for commercial and recreational fishing.[1][2]

4.4.1 References

[1] Hestelow, Andrew. "Deepwater Fishing Brown's Mountain" (PDF). *The Boat Mag*. Downrigger Shop. Retrieved 2013-12-13.

[2] "Map of Browns Mountain, NSW". Bonzle.com. Retrieved 2013-12-13.

4.5 Carondelet Reef

Main article: Kiribati

Carondelet Reef is a horseshoe-shaped reef of the Phoenix Islands, also known as the Rawaki Islands, in the Republic of Kiribati. It is located 106 km southeast of Nikumaroro, at 05°34′S 173°51′W / 5.567°S 173.850°W, and has a least depth of 1.8 meters. It is reported to be approximately 1.5 km in length.[1] The sea occasionally breaks over it.

4.5.1 History and Name

Carondelet Reef was named after the vessel which reported it, which was not the USS Carondelet. It had been previously referred to as Carondelet reef before the USS Carondelet was named and commissioned, as mentioned in a report of the air search for Amelia Earhart dated 16 July 1937.[2]

The multiple positions of Winslow Reef, mentioned by Robert Louis Stevenson, may have been due to confusing the position of Carondelet Reef with Winslow Reef.

4.5.2 Status

The reef is part of the Phoenix Islands Protected Area as an underwater nauture reserve.

4.5.3 See also

- List of Guano Island claims
- Winslow Reef, Phoenix Islands

4.5.4 References

[1] Mike Pearson; Jonathan Willis-Richards. "Islands of Kiribati". Archived from the original on 2006-12-31. Retrieved 2007-02-21.

[2] John Lambrecht (1937-07-16). "Lt. John Lambrecht's Report on the Search of the Phoenix Islands". *The Earhart Project*. TIGHAR. Retrieved 2007-02-21.

4.5.5   External links

- Phoenix Islands Protected Area, Kiribati

- Oceandots at the Wayback Machine (archived December 23, 2010)

Coordinates: 05°34′S 173°51′W / 5.567°S 173.850°W

4.6   Chelan Seamount

The **Chelan Seamount** is a seamount located in the Pacific Ocean off the coast of northern Vancouver Island, British Columbia, Canada.

4.6.1   References

4.6.2   See also

- Volcanism in Canada

- List of volcanoes in Canada

4.7   Cobb–Eickelberg Seamount chain

The **Cobb–Eickelberg Seamount chain** is a seamount chain stretching from the Aleutian Trench in the north to Axial Seamount, the youngest volcano in the chain, which lies approximately 480 kilometres (300 miles) west of Cannon Beach, Oregon. The chain was created by the Cobb hotspot as the Pacific Plate drifted in a northwesterly direction, leaving a trail of seamounts in its wake. The most recent volcanic activity in the chain was at Axial Seamount on January 28, 1998, when a magnitude 4.7 earthquake was recorded in the area.[1]

Some features in the chain:

- Cobb hotspot

- Axial Seamount

- Brown Bear Seamount

- Cobb Seamount

- Patton Seamount

4.7.1   References

[1] VolcanoWorld: Axial Seamount Retrieved 2007-09-14

4.8   Cordell Bank National Marine Sanctuary

Cordell Bank National Marine Sanctuary is a marine sanctuary located off the coast of California. It protects an area of 339 sq nmi (449 sq mi) of marine wildlife. The administrative center of the sanctuary is on an offshore granite outcrop 4.5 sq mi (12 km^2) by 9.5 sq mi (25 km^2), located on the continental shelf off of California. The outcrop is, at its closest (Point Reyes), 6 mi (10 km) from the sanctuary itself.[2]

Cordell Bank is one of the United States' 13 National Marine Sanctuaries that protect and preserve ocean ecosystems in the U.S. Cordell Bank is a seamount approximately 50 miles (80 km) northwest of San Francisco where the ocean bottom rises to within 120 feet (37 meters) of the surface. The seamount was discovered in 1853 by the U.S. Coast Survey, and named for Edward Cordell, who surveyed the area more thoroughly in 1869. It was extensively explored and described during 1978-86 by Robert Schmieder, who published a monograph about it [Schmieder, 1991]. It has been protected as a sanctuary since 1989. The protected area encompasses 526 square miles (1347 km^2) of ocean.

The unique blend of ocean conditions and undersea topography creates a rich and diverse underwater ecosystem. A subsurface island rises from soft sediments covering the continental shelf. The upper pinnacles reach to within 115 ft (35 m) of the surface, and the average depth is 400 ft (122 m). The sanctuary serves as a breeding ground for migratory marine mammals, birds, and fish. The prevailing California Current flows southward along to coast, causing an upwelling of nutrient-rich water that provided the foundation for the area's marine ecosystem.[2]

Sanctuary regulations prohibit extraction of hydrocarbons (oil, natural gas), the removal of benthic (bottom-dwelling) organisms, discharge of wastes, and removal of cultural resources. Recreational scuba diving is not recommended in the sanctuary due to depth and currents.

4.8.1 Geological setting

Cordell Bank was originally created 93 million years ago, as a member of the Sierra Nevada. The grinding of the plates at the San Andreas Fault, with the Pacific Plate moving north and the North American Plate moving south, parts of the Sierra Mountains were sheared off and carried northwards, including Cordell Bank. Eventually this grinding carried Cordell Bank to its present location opposite Reyes Point. Cordell Bank is still moving, by an average of 3 cm (1 in) per year.[3]

Between 20,000 and 15,000 years ago the sea level in the area was 360 ft (110 m) below the current level, leaving most of Cordell Bank exposed and making it a true island. Today the bank rises out of soft sediment, deposited on the bank more recently by coastal erosion. Within just 7 miles (11,265 m) of Cordell Bank, the continental shelf drops to over 1 mi (2 km) deep.[3] The seamount is largely composed of granite.

4.8.2 History

Coastal California has a rich history of marine utilization by Native Americans and early settlers. Cordell Bank was a mystery prior to the 19th century because neither the Miwok natives nor the settlers had any incentive to venture far out from shore, when food resources were available close to shore. Many European mariners sailed right over Cordell Bank without even knowing it was there.[4]

In the later half of the 1800s there was a strong incentive to survey the coast of California so as to promote maritime safety. Cordell Bank was discovered in 1853 by George Davidson of the US Coast Survey during a mapping expedition on California's north coast.[4][5]

In 1869 Edward Cordell (the reserve's namesake) was sent to collect additional information on a "shoal west of Point Reyes". He found the area by following the numerous birds and maritime mammals. To measure the depth, Cordell lowered a lead weight into the water until it hit the bottom and then measured the length of the line on its return to the surface.[5] The area was considered a productive fishing area, but not much was discovered about its marine life until an expedition in 1977.[4]

NOAA carried out a detailed multibeam echosounder survey of the area in 1985 from aboard the R/V *Davidson*.[5]

The expedition was led by a non-profit research group, Cordell Expeditions.[5] Over the next 10 years, scores of underwater dives documented the organism and wildlife living in, on, and around the bank. The efforts gave rise to an understanding of the diversity in the back, and the efforts were instrumental in the decision to make it a sanctuary.[4]

Expansion

A bill (H.R. 5352) was proposed to Congress by Representative Lynn Woolsey to expand the size of the Cordell Bank National Marine Sanctuary and the neighboring Gulf of the Farallones National Marine Sanctuary by 1,094 square miles (2,830 km^2).*[6]

In 2012, the National Oceanic and Atmospheric Administration proposed an expansion of the borders of the sanctuary, along with expansion of the Gulf of the Farallones National Marine Sanctuary, to include an additional 2,700 square miles, reaching to Point Arena.*[7]

Expansion was passed March 2015.*[8]

4.8.3   Biology

26 species of marine mammals, including whales, dolphins, seals, and sea lions, are known to frequent the waters of the sanctuary. In addition, Cordell Bank is one of the most important feeding grounds in the world for the endangered blue and humpback whales; these species travel all the way from their breeding grounds in coastal Mexico and Central America to feed on the krill that aggregate near the bank. Another unique species is the Pacific white-sided dolphin (*Lagenorhynchus obliquidens*), which can be seen in large numbers. Other visitors include California sea lions (*Zalophus californianus*), Northern elephant seals (*Mirounga angustirostris*), Northern fur seals (*Callorhinus ursinus*), and steller sea lions (*Eumetopias jubatus*), all of which are attracted to the abundance in krill, squid, and juvenile fish.*[9] Leatherback Sea Turtles also inhabit sanctuary waters.

Cordell Bank is also a major foraging ground for passing seabirds. Known as the "Albatross capital of the world," 5 of the 14 major species of albatross have been documented there. The two most common are the Black-footed albatross (*Phoebastria nigripes*) and Sooty shearwater (*Puffinus griseus*). It is also one of the few places to see a Short-tailed albatross (*Phoebastria albatrus*), which is extremely rare; the species was thought to have gone extinct after World War II. Currently the world population hovers at around 1000 individuals.*[9]

Cordell Bank is also known for its abundance of fish. Flatfish, most notably sanddabs, make their home on the mud of the seafloor in the sanctuary. Both solitary and schooling fish find refuge for predators among the bank's rocky pinnacles. Cordell Bank supports more than 246 species of fish, including 44 species of rockfish, ranging in size from the 8-inch Pygmy rockfish to the 3-foot (0.91 m) Yelloweye rockfish.*[9]

Although far from shore, sport fishers prize Cordell Bank as a fishing spot, and regularly venture out from shore to catch albacore and salmon.*[9]

Ecologic cycle

This unusual granite mountain is surrounded on three sides by deep waters, which allows the flow of deep nutrient-rich waters over relatively shallow waters with sufficient light to support photosynthesis.

The ecological cycle at Cordell Bank can be divided into three oceanographic seasons. During the spring, strong northwestern winds push the water southward along the California coast. Gale winds and the Earth's rotation drive surface water away from the shore, only to be replaced by an upwelling of colorer, deeper, and more nutrient-rich waters from offshore. this contributes to the growth in numbers of phytoplankton, which are the foundation of the marine food web, in turn leading to a rise in food, and thus numbers, of the organism higher on the chain.*[10]

During the late summer and fall seasons, the coastal winds that stirred up the deeper waters die down, and the northward-flowing Davidson Current prevails, bringing warm but nutrient-poor water from the south.*[10]

During the winter storm months, the sea is dominated by rough weather, which mixes the deeper water with that above. The temperature on top of the continental shelf mixes, and the temperature, salinity, and concentration of water and the nutrients in them assimilate.*[10]

4.8.4 References

[1] U.S. Geological Survey Geographic Names Information System: Cordell Bank National Marine Sanctuary

[2] "About the Sanctuary". NOAA. Retrieved 25 November 2009.

[3] "Natural Natural Environment: Geologic History". NOAA. Retrieved 26 November 2009.

[4] "Natural The History of Cordell Bank". NOAA. Retrieved 26 November 2009.

[5] Schmieder, RW (1985). "The expeditions to Cordell Bank". *In: Mitchell, CT (ed). Diving for Science⋯1985. Proceedings of the Joint American Academy of Underwater Sciences and Confédération Mondiale des Activités Subaquatiques annual scientific diving symposium 31 October - 3 November 1985 La Jolla, California, USA.* Retrieved 2013-05-27.

[6] Woolsey, Lynn C (2004-10-08). "H.R.5352 - Gulf of the Farallones and Cordell Bank National Marine Sanctuaries Boundary Modification and Protection Act". Congressional Bills 108th Congress: From the United States Government Printing Office. Retrieved 2013-05-27.

[7]

[8]

[9] "Natural Environment: Biological Resources". NOAA. Retrieved 26 November 2009.

[10] "Natural Environment: Oceanographic Seasons". NOAA. Retrieved 26 November 2009.

Works cited

- "A National Marine Sanctuary: Cordell Carpenter Bank," Published by the National Marine Sanctuary Program of the National Oceanic and Atmospheric Administration (NOAA)

- Schmieder, Robert W., 1991, *Ecology of an Underwater Island*, Published by Cordell Expeditions, Walnut Creek, CA, http://www.cordell.org. See also: Schmieder Bank

- Stallcup, Richard. 1990. **Ocean Birds of the Nearshore Pacific**, Point Reyes Bird Observatory, Stinson Beach, CA. 214 pp.

4.8.5 External links

- Official sanctuary site

- Black-footed albatros clip

- Cordell expeditions

- Cordell Marine Sanctuary Foundation

4.9 Cortes Bank

Cortes Bank is a shallow seamount (a barely submerged island) in the North Pacific Ocean. It is 96 miles southwest of San Pedro, Los Angeles, 111 miles (166 kilometers) west of Point Loma San Diego, USA, and 47 miles (82 kilometers) south-west of San Clemente Island in Los Angeles County. It is considered the outermost feature in California's Channel Islands chain. At various times during geologic history, the bank has been an island, depending on sea level rise and fall. The last time it was a substantial island was around 10,000 years ago during the last ice age. It is quite possible that this island was visited by the first human inhabitants of the Channel Islands, most notably San Clemente Island, whose seafaring residents would have been able to see "Cortes Island" from high elevations on clear days.[1]

The shallower reaches of the bank comprise about 15–18 miles of sandstone and basalt and they rise from the ocean floor from 1000 fathoms, or just over a mile in depth. The bank has been described as a series of mountaintops, but really it is more of the shape of a wave-scoured mesa with a few hard, basaltic high spots along its length. The shallowest peak, the **Bishop Rock**, rises to between 3 and 6 feet (1–2 m) from the surface, depending on the tides. On very low tides, the rock can be visible in the trough of passing waves. Other shoal spots besides the Bishop Rock also spawn giant waves. These shoals range in depth from 30 to 100 feet and are a hazard to shipping. Nine Fathom spot is about 4.5 miles (7 kilometers) northwest of Bishop Rock and also rises to about 54 feet (18 m) below the surface. Both are noted scuba diving locations featuring clear water, vast kelp forests and abundant sea life. The Bishop Rock also creates a renowned big-wave surfing spot recognized as being capable of producing some of the tallest surfable waves in the world.*[2]*[3]*[4]

4.9.1   General

It has long been reported that the Cortes Bank was discovered in modern times by the captain of the side-wheel steamship *Cortes*, TP Cropper. In 1853, during a voyage from Panama to San Francisco, Cropper reported seeing the seas "in violent commotion" above an uncharted seamount that would eventually be named after the ship. Cropper at first thought he was above a volcano.*[5] However, it seems likely that the very first modern sighting of the Bank was not by Cropper but by US Navy Lt. James Alden and Captain Jonathan "Mad Jack" Percival. This occurred on January 5, 1846. At that time, the frigate *USS Constitution* was passing well off the US West Coast from Monterey to see duty in the Mexican American War. The logbook of the *Constitution* from this day puts the ship in the vicinity of the bank and reads: "At 4-20 (p.m.) discovered breakers bearing N.E. about 10 miles distant." *[6]

Alden would eventually become an officer with the United States Coast Survey, an organization charged with mapping the U.S. coastline. In the wake of the Cortes sighting, and because of his own earlier sighting, Alden dispatched the crew of the *USS Ewing* to discover the source of the open ocean breakers. Under Alden's orders, Lt. TH Stevens discovered and mapped the location and a rough outline of the Bank, which was for years incorrectly named "Cortez Bank." Stevens discovered waters around 54 feet deep, although he failed to discover the dangerously shallow area around the Bishop Rock, and it does not show up on the first Coast Survey map published in 1853.*[7]

Bishop Rock is today marked by a nearby warning buoy. It was named for the clipper ship *Stillwell S. Bishop* that reportedly struck the rock in 1855, then continued to San Francisco with a patched hull. There is some uncertainty over whether the *Bishop* actually struck the rock, though the captain of the ship, William Shankland, surely at least encountered waves along its periphery, likely in 1854. In the wake of the *Bishop*'s voyage, James Alden placed a talented navigator and inveterate explorer from Wilmington, NC named Lt. Archibald MacRae, USN in command of the *Ewing* and dispatched him to discover the Bank's shoalest reach. On November 3, 1855, the *New York Times* carried the story "Dangerous Rock off the Coast of California," which reported MacRae's finding and the fact that he and the crew of the ship had anchored a pair of casks bearing a flag to mark the spot. Two weeks after the story appeared, MacRae committed suicide aboard the *Ewing* in San Francisco Bay, shooting himself in the head with a large caliber Colt revolver, while anchored alongside Alden's ship, the USS *Active*.*[8]*[9]

Among other notable events in the history of the Cortes Bank is the fairly disastrous exploration of the Bank for treasure in 1957 by Mel Fisher. He was convinced that the wreckage of a Spanish Galleon lay on the seafloor off the Bishop Rock. The expedition found no treasure, but the ship carrying Fisher burned nearly to her waterline.*[10]

There have been at least two efforts to turn the Cortes Bank into an island nation. The most notable occurred in late 1966, when a team of entrepreneurs planned to turn the Cortes Bank into the constitutional monarchy of **Abalonia**. The general plan was to scuttle a WWII era concrete hulled freighter—probably the Tampa-built McClosky ship *Richard Lewis Humphrey*, which was later badged *Jalisco* in Mexico*[11]—atop the Bishop Rock in very shallow water and surround the ship with an ever expanding ring of boulders so she could be used as a seafood processing factory. The group reasoned that international maritime law would allow them to become the rulers of their own nation because the Bank lay in international waters.*[12] The ship was instead destroyed atop the Bishop Rock by the same waves that are surfed today and her crew was nearly killed. The wreck of the *Jalisco* today lies beneath the surf zone in three pieces in 6 to 40 feet (2–12 m) of water, and is a diving location.*[13]*[14]

When another company planned to form a nation called **Taluga**, the US government declared that the bank, as part of the continental shelf, was US territory.*[15]*[16]

On 2 November 1985 the aircraft carrier USS *Enterprise* (CVN-65) struck the Cortes Bank reef about one mile east of Bishop Rock, putting a 60-foot (18 m) gash in her outer hull on the port side, ripped-off her port keel, and severely deformed her outboard port propeller blades. She continued operations, then went into dry dock at Hunter's Point Shipyard for repairs.[17][18]

4.9.2 Surfing

In the summer of 1961, a surfer named Harrison Ealey of Oceanside, California became one of the very first people to surf a wave at the Bishop Rock. In around 1973, surfer Ilima Kalama, father of famed big wave surfer Dave Kalama, nearly lost his life when the abalone fishing boat he was aboard sank on the Bishop Rock in the middle of the night.

In the early 1990s Larry Moore, photo editor at *Surfing* magazine, and Mike Castillo, veteran surfer and pilot, made flights out across the bank on rumors of giant waves. During a monster swell in 1990 that has been dubbed "The Eddie Aikau Swell," they were astonished when they found empty waves breaking atop the bank in the 80 to 90 foot range. By 1995 Moore had seen and photographed waves and that year he led an expedition with a small group of surfers out there (including *Surfing* magazine editors Sam George and Bill Sharp) and pro surfer George Hulse. The team found relatively small but very glassy waves in the fifteen foot range, and George Hulse was the first to catch one. "It was the only time I wrote out a will before a surf trip," Sharp said of the mission.[19][20]

Several surfers planned for the ideal conditions at the bank. In 2001 a storm called "Storm 15" in the Gulf of Alaska and a high pressure ridge over California came together to create huge swells but light wind over the bank. A team of surfers went out on the F/V *Pacific Quest* from San Diego, with big-wave tow surfers Ken Collins, Peter Mel, Brad Gerlach and Mike Parsons, plus paddle-surfers Evan Slater and John Walla. On the morning of 19 January 2001 they found smooth glassy conditions and enormous, half-mile long waves breaking across about 1 mile (1.5 kilometer) of reef. Walla and Slater tried to paddle for one of these waves and both nearly drowned.

Larry Moore photographed from a circling plane, Dana Brown shot from a boat for his surf film *Step Into Liquid*, and Fran Battaglia shot from two other boats for his wave science film *Making The Call: Big Waves of the North Pacific*, his documentary for Swell, XXL, NBC Dateline, *Billabong Odyssey* and Activision's Kelly Slater Pro Surfer video game. Parsons was towed into the wave of the day. His very first ride at the Cortes Bank was estimated at 66 feet (20 m) on the face. It won him the first of two Guinness World Records and the Swell XXL Biggest Wave Award (now Billabong XXL) prize of $66,000 for the biggest wave surfed in 2000/2001.[21][22][23]

On January 5, 2008, Mike Parsons, Brad Gerlach, Grant "Twiggy" Baker and Greg Long returned to the location in the midst of one of the worst storms ever recorded off the coast of California. Mike Parsons was photographed on a wave bigger than his award-winning ride of 2001, judged by the Billabong XXL judges as 70+ feet on the face—later determined to be at least 77 feet—and Parsons second Guinness World Record.[24][25][26][27][28]

Although remote, the Cortes Bank draws crowds when conditions are good. On a trip with the Billabong Odyssey in January 2004 Sean Collins counted 10 or 12 boats with about 40 surfers.

4.9.3 See also

Surf films featuring Cortes Bank:

- *Billabong Odyssey*

- *Step Into Liquid*

- *Ghost Wave*

4.9.4 References

[1] Porcasi; Judith and Paul (1999). "Early Holocene Coastlines of the California Bight". *Pacific Coast Archaeological Society Quarterly*. 2 **35** (Spring/Summer).

[2] Dixon, Chris (2011). *Ghost Wave*. San Francisco, CA: Chronicle Books. pp. 19–28. ISBN 978-0-8118-7628-5.

[3] Holzman, J.E. (1952). "The Submarine Geology of Cortes and Tanner Banks". *Journal of Sedimentary Research* **22**.

[4] Raab, Cassidy, and Yatsko, D. L. Mark, Jim, and Andrew (2009). *California Maritime Archaeology: A San Clemente Island Perspective*. Lanham, MD: Alta Mira Press. ISBN 0759113165.

[5] *Report to the Superintendent of the Coast Survey Showing the Progress of the Survey During the Year 1853,1854, 1855*. Washington, DC: United States Coast Survey. 1853, 1854, 1855. Check date values in: |date= (help)

[6] Dixon, Chris (2011). *Ghost Wave*. San Francisco, CA: Chronicle Books. p. 245. ISBN 978-0-8118-7628-5.

[7] *Report to the Superintendent of the Coast Survey Showing the Progress of the Survey During the Year 1853,1854, 1855*. Washington, DC: United States Coast Survey. 1853, 1854, 1855. Check date values in: |date= (help)

[8] Theberge, Albert E. (Spring–Summer 2006). "Charting Our Destiny". *Mains'l Haul, a Journal of Pacific Maritime History*.

[9] Editors (Summer 1968). "About a Rock—and a Bishop" (PDF). *Mains'l Haul: A Journal of Pacific Maritime History* **5** (Summer): 3.

[10] Beronius, George (January 14, 1957). "Sunken Treasure! Shout Lures 23 on Sea Search". *Los Angeles Times*.

[11] Bender, Rob. "Editor, www.concreteships.org".

[12] Stewart, Hal D (October 31, 1966). "Pair Planning Island Nation off San Diego". *The Pasadena Independent*.

[13] Keen, Harold (November 17, 1966). "Promoters of Abalone Ship Plan May Face Federal Prosecution". *The Los Angeles Times*.

[14] Dixon, Chris (2011). *Ghost Wave*. San Francisco, CA: Chronicle Books. pp. 58–76. ISBN 978-0-8118-7628-5.

[15] Cortez Development Corporation (1966). *A Plan for An Island State*. Missing or empty |title= (help)

[16] James L. Erwin, *Atlas of Forgotten Nations*, quoted in A Shoal Less-Traveled... Until Now by Michael Kew

[17] USS Enterprise (CVN 65) page

[18] Roberts, Karlene H. (1990). "Bishop Rock Dead Ahead: The Grounding of the USS Enterprise". *Submitted to: The US Naval Institute's Proceedings - but not published in the journal*.

[19] Slater, Evan (June 2001). "Project Neptune". *Surfing Magazine*.

[20] Dixon, Chris (2011). *Ghost Wave*. San Francisco, CA: Chronicle Books. pp. 8–18. ISBN 978-0-8118-7628-5.

[21] Slater, Evan. pulse/2001/jan/01_22_bank_one.cfm "Into Thick Water Part One: Mike Parsons, Brad Gerlach, Peter Mel and Ken 'Skin- dog' Collins summit Cortes Bank" Check |url= scheme (help). Surfline.com.

[22] Slater, Evan. mag/pulse/2001/jan/01_22_bank_two.cfm "Into Thick Water: Part Two" Check |url= scheme (help). Surfline.com.

[23] Slater, Evan (June 2001). "Project Neptune". *Surfing Magazine*.

[24] Dixon, Chris (Jan 9, 2008). "Surfers Defy Giant Waves Awakened by Storm,". *The New York Times*.

[25] Downes, Lawrence (January 11, 2008). "The Next Sir Edmund Hillarys: Riders on the Storm". *The New York Times*.

[26] Parsons, Mike. "Eye of the Storm. Gerlach, Snips, Greg Long and Twiggy Score Huge at Cortes Bank". Surfline.com. Retrieved Jan 8, 2008.

[27] Casey, Susan, "*Reef Madness*", Sports Illustrated, Vol. 108, No. 2, pages 50-52, 2008, January 21,

[28] Dixon, Chris (2011). *Ghost Wave*. San Francisco, CA: Chronicle Books. pp. 200–232. ISBN 978-0-8118-7628-5.

4.9.5 Links

- Chris Dixon, "Ghost Wave," The Discovery of Cortes Bank and the Biggest Wave on Earth," 2011, by Chronicle Books, San Francisco, CA. ISBN 978-0-8118-7628-5, http://ghostwavebook.com

- Samuel Pyeatt Menefee, "Republics of the Reefs": Nation-Building on the Continental Shelf and in the World's Oceans, *California Western International Law Journal*, vol. 25, no. 1, Fall, 1994, pp. 102–03.

- Big-Wave Surfing at Cortes Bank by Jack Boulware 10/6/2004 (also in Southwest Spirit magazine)

- Surfing Tsunami by Dr Karl Kruszelnicki at the Australian Broadcasting Corporation

4.10   Cross Seamount

Cross Seamount is a seamount far southwest of the Hawaii archipelago, about equidistant from the cities of Honolulu and Kona. It is one of the numerous seamounts surrounding Hawaii, although unrelated to the Hawaiian hotspot.[2] It is notable for being one of the best studied of the numerous seamounts surrounding Hawaii, as it has been included in numerous biological surveys, most recently in 2007. It is also a site of offshore fishing, for its abundant tuna. The fishery management problems at Cross Seamount are typical of management problems in many fisheries, and its small size makes it a scientifically useful model for the analysis of fishery management.[3]

4.10.1   Geology

Cross Seamount, a landform arising from the ocean floor, is one of the more distant lesser seamounts surrounding the island of Hawaii.[2] The NOAA has taken advantage of Cross Seamount's position, and planted several weather buoys on the seamount.[4] Very little is known about Cross Seamount's geology, as all of its studies as of March 2009 were of its ecology; however, it was mapped in 1996 by a four year effort to map all of Hawaii's seafloor in detail.[2]

4.10.2   Biology

History of fishing

Cross Seamount is well known in fishery and scientific communities for its small but abundant biologic community. Longline fishing ships occasionally fished at Cross Seamount for many years prior to the development of handline fishing, in 1976. The level of fishing remained fairly constant up until the mid 1980. In the late 1980s, the amount of fishing activity at Cross Seamount more than quadrupled; however, it is hard to tell because of limited data.[3]

Local tuna populations belong to the same general stocks of tuna that are widely distributed throughout the Pacific.[3] There are no accurate size data collected on the catches from fishing, aside from limited port sampling by the National Marine Fisheries Service. However, it is known that the largest percent of the catches are juvenile tuna between 10 and 35 pounds.[3] These fish are probably somewhere between one and two years of age, well below sexual maturity. Fishermen report that the offshore fishing grounds are productive year-round. However, during summer months, large Yellowfin tuna tuna are more abundant. The weight of the catches are estimated to be about 75% *Thunnus obesus* (Bigeye tuna), and 25% *T. albacares* (Yellowfin tuna).[3] According to estimates, about 1 million pounds of Bigeye tuna and 400,000 pounds of Yellowfin were fished from offshore operations from the region in 1995.[3] The amount of tuna is thought to depend on conditions on the seamount, and on the current process range of the tuna schools.[5]

Concerns and studies

Reviews of the local catches have seemed to show no impact of the fishing operation on nearby tuna fishing. However, this is not very surprising, as fishery operations exist nearby with as much as 1000 times the activity. Nevertheless, as tuna spawn locally, there are concerns that the fishing eats away at the juvenile tuna population, which has the potential to seriously hurt fish numbers.*[3]

1995 tagging project   In 1995, a project to tag the local fish, to analyze the population, was implemented by the NOAA, based on a plan proposed by fishermen in 1992.*[3] The first fish were tagged and released in August 1995.*[3] The objectives of the project was to investigate the retention rates of tuna on the Cross Seamount, movement patterns of tuna in relation to the fishery patterns, and the interaction between surface and longline fishing.*[4]

The recovery rate was found to be 5.2% for Yellowfins, and 8.1% for Bigeyes.*[4] The longest-distance recapture was a yellowfin tuna tagged on the Cross Seamount in November 1996 and recaptured off the coast of the Baja California peninsula in July 1997.*[4]

Overfishing (as in the rate of depletion) at the Cross seamount appeared to be quite high, causing concerns. However, as much of the population seems to originate elsewhere, this loss is most likely accounted for.*[3] Problems with fishery interaction were eliminated, as 90% of the recaptures were in the local area. However, it was not possible to create a model on the overfishing of juveniles, as there was insufficient data to predict this accurately.*[3]

HARP   Cross seamount hosts a High-frequency Autonomous Acoustic Recording Package (HARP), installed on its summit by the NOAA in April 2005.*[6] The purpose of the HARP is to monitor cetaceans that near the seamount, and ultimately to develop ways to work with the marine mammal population using acoustic data. The unit undergoes routine maintenance, and its data is retrieved and analyzed every so often by an NOAA ship.*[7]

NOAA cruises   Cross Seamount was cruised by the NOAA in 2006, 2007, and 2008.

On November 2, 2006, scientists on the NOAA Ship *Oscar Elton Sette*, led by chief scientist Michael Musyl, conducted an array of biological experiments off of Kona coast and Cross Seamount. Their goal was to investigate and find ways to reduce the impacts of fishing and improve knowledge of the distribution of tunas, and how they are impacted by fishing activity.*[7] It also performed routine maintenance on the HARP.*[7]

On April 21, 2007, another cruise embarked for Cross Seamount. This cruise's goal, led by Reka Domokos and again on the *Oscar Elton Settle*, was to test and develop new methods for estimating fish populations using bioacoustics, specifically that of the bigeye tuna.*[8] The study was also keyed to try to determine the relationship between seamounts and the local fish population.

On April 15, 2008, another cruise embarked, to study the bigeye tuna population and their migratory patterns. The researchers, again on the ship *Oscar Elton Sette*, investigated the seamount waters, measuring the number of bigeye tuna and the prey species that frequent the area.*[5]

4.10.3   References

[1]   "Detailed Seamount Information - Cross Seamount" . *Seamounts Database*. Retrieved 2009-03-26.

[2]   "Hawaii's Volcanoes Revealed" (PDF). *USGS Poster*. USGS. Retrieved 2009-03-28.

[3]   Sibert, John; Kim Holland; David Itano (December 4, 1997). "Tuna Fishing at Cross Seamount" . *Slideshow Presentation*. Pelagic Fishes Research Program, JIMAR. p. 29. Retrieved 2009-03-27.

[4]   Holland, Kim (1997). "A Tag and Release Program for the Hawaiian Seamount Yellowfin and Bigeye Tuna Handline and Troll Fisheries" . Retrieved 2009-03-27.

[5]   "Scientists using underwater acoustics to study bigeye tuna and their prey at Cross Seamount" . NOAA. April 2008. Retrieved 2009-03-27.

[6]   "Cruise report" (PDF). *NOAA - Cruise Confirmation*. NOAA. April 2005. Retrieved 2009-03-27.

[7]   "Scientists on NOAA Research Cruise are Studying the Physiology of Pelagic Fishes and Ways to Reduce Impacts of Longline Fishery Bycatch" . NOAA. November 2006. Retrieved 2009-03-27.

[8] "Scientists on the NOAA Ship Oscar Elton Sette are conducting a bioacoustics survey to learn about the distribution and abundance of bigeye tuna and its prey at Cross Seamount" . NOAA. April 2007. Retrieved 2009-03-27.

4.11   Davidson Seamount

Davidson Seamount is a seamount (underwater volcano) located off the coast of Central California, 80 mi (129 km) southwest of Monterey and 75 mi (121 km) west of San Simeon. At 26 mi (42 km) long and 8 mi (13 km) wide, it is one of the largest known seamounts in the world.*[4] From base to crest, the seamount is 7,480 ft (2,280 m) tall, yet its summit is still 4,101 ft (1,250 m) below the sea surface. The seamount is biologically diverse, with 237 species and 27 types of deep-sea coral having been identified.*[1]

Discovered during the mapping of California's coast in 1933, Davidson Seamount is named after geographer George Davidson of the U.S. National Geodetic Survey. Studied only sparsely for decades, NOAA expeditions to the seamount in 2002 and 2006 cast light upon its unique deep-sea coral ecosystem. Davidson Seamount is populated by a dense population of large, ancient corals, some of which are over 100 years of age. The data gathered during the studies fueled the making of Davidson Seamount into a part of the Monterey Bay National Marine Sanctuary in 2009.

4.11.1   Geology

A seamount such as Davidson is an underwater volcano; this one rises over 3,280 ft (1,000 m) above the surrounding ocean floor. Although there are over 30,000 seamounts in the Pacific Ocean alone, only about 0.1% of them have been explored.*[4] The aqueous environment of the seamount means that it behaves differently from volcanoes on land. Its surface is composed mostly of blocky lava flows, although some pillow lava, which is the typical lava type of a seamount, prevails at the deeper flank. The summit is composed of layered deposits of volcanic ash and pyroclastic material. These rocks indicate mildly explosive eruptions of gas-rich lava near the summit of the volcano. The base of Davidson is probably buried in a deep layer of muds.*[3]

At 26 mi (42 km) long and 8 mi (13 km) wide, Davidson Seamount is impressively large. If it were on land, it would dominate the landscape in a way similar to how Mount Shasta dominates the horizon of northern California. Put in perspective, the size of the seamount is enough to fill Monterey Bay from the Santa Cruz boardwalk to Monterey's Fishermen's Wharf.*[5]

Davidson Seamount is part of a group of seamounts off the continental margin, including Guide, Pioneer, Gumdrop, and Rodriguez seamounts, all located roughly between 37.5° and 34.0° degrees of latitude. This group is morphologically unique, and very similar to one another. All the seamounts in the group are complex northeast-southwest trending structures, consisting of parallel ridges separated by sediment-filled troughs. The ridges constructed run parallel to an ancient spreading center which has since been replaced in its role by the San Andreas Fault system.*[3] They are unique in this origin, as they are formed from the remnants of an old ocean-ridge spreading center.*[6] A series of "knobs" are aligned with the ridges; however the distinctive summit crater, evident in many oceanic volcanoes, is absent. This lack of a collapse crater suggests that magma was never stored in a chamber within the structure, as with most other volcanoes.*[3]

Analysis of argon–argon dating studies indicate that Davidson formed between 9 and 15 million years ago, 5 to 12 million years after the formation of the overlaying oceanic crust.*[3]

4.11.2 Ecology

Benthoctopus sp. and a clam near the summit of Davidson Seamount, 1,461 m (4,793 ft) deep.

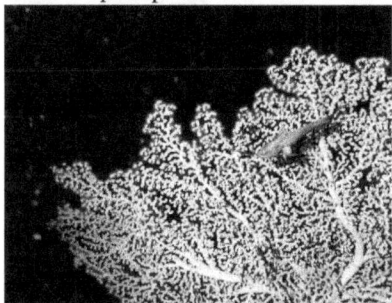

Grenadier fish (*Coryphaenoides sp.*) in front of bubblegum coral (*Paragorgia arborea*) on the crest of Davidson Seamount, at 1,255 m (4,117 ft) deep. Paragorgia grow to over 2.5 m (8 ft) tall on Davidson; they also grow in Monterey Canyon, but are generally smaller and less abundant.[7]

Studies have indicated that a seamount functions as an "oasis of life," with a higher species count and more biodiversity then the surrounding seafloor.[4] Although previous analysis has stressed the exceptionally of the seamount habitat, recent biological analysis, much of it centered on Davidson Seamount, has indicated that this does not necessarily translate into a higher endemic percentage.[8][7][9] However, it *is* believed that they provide a refuge for rare species that have difficulty surviving elsewhere.[8]

There are reasons that seamounts are biologically important. They rise high in the water column, creating complex current patterns that support life on, around, and above the seamount. The surface of the seamount also provides a substrate upon which organisms can attach themselves and grow. This in turn supports the species that feed on them, in turn supporting the whole ecological food web. Scientists have found that seamounts often provide a habitat for endemic species that are not found anywhere else.[4]

Davidson Seamount is among the best biologically described seamounts worldwide. Six major expeditions to the volcano have yielded over 60,000 species observations. As of 2009 scientists have observed and recorded 168 species of megafauna on the seamount. Of these, about 7% of the species at Davidson are endemic, meaning they live only at Davidson. 71% of the species can be confidently classified as cosmopolitan (widespread), and sufficient data exists for 22% of the observed species to strongly suggest that their range is not limited to seamounts. The remaining 7% have only been seen in video footage. Interestingly, 13 species have been identified in other areas, but never in other seamounts.[8]

The seamount is populated by a large variety of deep-sea corals, most of which in turn provide a habitat for other species. It has in the past been called "An Oasis in the Deep", hosting a vast coral forest, large sponge field, crabs, deep-sea fish, basket stars, and a number of rare benthic species, some of which have yet to be studied properly or even named.[1] These are all cold-water species, as the temperature even at the top of the seamount is just above freezing—around 2 °C (36 °F), as compared to 14 °C (57 °F) at the ocean's surface.[6]

Despite its size, the top of Davidson Seamount remains over 4,500 ft (1,372 m) below the ocean's surface. This great depth means that the habitats that the seamount supports have not been significantly disturbed by human activity; anchoring and trawling typically does not occur below a depth of 1,500 ft (457 m), and waste disposal and discharge occurs much closer to shore.[5]

In comparisons drawn to the nearby Monterey Bay National Marine Sanctuary, the two were shown to be very different biologically. Species that are rare in Monterey Bay are common on Davidson, and vice versa.[*][8]

Deep-sea coral

The ecology on Davidson Seamount is dominated by an extensive "forest" of ancient deep-sea coral and sponges, some of which are over 100 years old. Although these species also grow elsewhere, they are generally sparsely distributed and far smaller and younger than the coral growth on Davidson. Conversely, sea cucumbers, which are very common on the walls of Monterey Canyon, are all but absent on Davidson, an example of the polarity between Davidson and Monterey Bay.[*][7]

Researchers speculate that Davidson Seamount is an ideal habitat for deep-sea corals and sponges because it has favorable bottom rock (bare lava rock), a steady food supply (supplied by a water current favoring the seamount; see the section above), and has not been disturbed by strong bottom currents than often bother other seafloor areas.[*][7]

Craig McClain, one of the scientists studying the seamount, told PhysOrg:

Research also suggests that seamounts such as Davidson Seamount may be ecologically valuable to rare species that use them as breeding grounds. The seamounts are likely to be a source of larvae that maintains the population of the species in nearby, sub-optimal areas, known as "sinks." Sinks are low-lying areas in which species can live, but do so very poorly; if they are not replenished by nearby population centers, such as at Davidson, they could disappear from the area entirely. DNA studies may in the future help scientists prove seamounts are indeed sources of larvae for their surrounding seafloor.[*][7]

4.11.3 History

Exploration

Davidson Seamount was initially discovered and mapped in 1933.[*][4] Davidson Seamount was the first underwater volcano to be classified as "seamount" by the United States Board of Geographic Names, in 1938, and was named in honor of the United States Coast and Geodetic Survey scientist George Davidson, one of the key figures in the survey of America's west coast.[*][1]

Because of its great depth, for a long time Davidson Seamount had been preceded only by a sparse few investigations. Davidson is interesting to volcanologists because of its unique geology, and to biologists for its unique ecology. In 2002 the NOAA sponsored the first modern in-depth study of the seamount. The team included scientists, educators, and resource managers, with the goal of documenting species, taking geologic samples, and describing the ocean environment.[*][10] The expedition documented many rare, previously undiscovered species that exist nowhere else, not even on nearby seamounts,[*][4] including ancient coral gardens that are vulnerable to human activity.[*][11]

Recent expeditions to Davidson have focused on its ecology, and specifically on the variety of deep-sea corals, some over 100 years old, that live on its banks. These large colonies are extremely fragile to human interaction. Davidson's proximity to scientific research institutions has helped its exploration, as multiple dives, mappings, and studies have made it one of the better-studied seamounts in the world.[*][4]

"We were blown away by the size, age, and diversity of deep-water corals we saw during the 2002 Davidson. Indeed, the discoveries we made during that cruise prompted members of the public to propose Davidson Seamount be protected as part of the Monterey Bay National Marine Sanctuary. We wanted to go back to learn why so many extraordinary corals thrive there and to determine their age and growth patterns." [*][12]

—Andrew DeVogelaere, chief scientist of the 2006 expedition.

In 2006, another exploration, a collaboration of the Monterey Bay National Marine Sanctuary, the Monterey Bay Aquarium Research Institute (MBARI), and Moss Landing Marine Laboratories was undertaken, mainly to test a model that had been developed to predict the availability of coral and to advance the understanding of the seamount's deep-sea coral.[*][11] The NOAA outlined a set of 4 goals for the expedition:[*][13]

- Understand why deep-sea corals live where they do on the seamount

- Determine the age and growth patterns of the bamboo coral

- Improve the species list and taxonomy of corals from the seamount

- Share the exciting experience with the public through television and the Internet

Scientific data on the water currents and food availability of the seamount was collected, as was information on the age and growth patterns of the corals themselves. The British Broadcasting Corporation (BBC) filmed the cruise for their series, *Planet Earth.*[11] A total of 70 hours of observations and 102 deep-sea animal and rock specimens were collected during the cruise.*[13] The expedition, which lasted from January 26 through February 4, made use if the MBARI's research vessel *Western Flyer* and the ROV *Tiburon.*[12]

As a marine sanctuary

Following the information learned from the 2002 and 2006 expeditions, there was public support for the making of Davidson Seamount into a marine sanctuary.*[12] A key group of research scientists, fishermen, officials, educators, and marine biologists was formed in 2006 to discuss whether or not to make Davidson Seamount a National Marine Sanctuary under the National Marine Sanctuaries Act, and if so whether to make it its own sanctuary, or incorporate it into the nearby Monterey Bay National Marine Sanctuary. The group decided that the seamount was eligible for sanctuary designation, and the majority of the group recommended its incorporation into the nearby Monterey Bay sanctuary.*[4]

The Monterey Bay advisory group concurred with the findings, and submitted its approval to the NOAA, with whom the final decision rested.*[4] In a press release dated November 20, 2008, the NOAA finalized its approval of the plans, and expanded Monterey Bay by a total of 775 sq mi (2,007 km^2) to include Davidson Seamount in its protected area.*[14] After a management plan was created, Davidson Seamount was incorporated into Monterey Bay in 2009, 7 years after it was originally proposed.*[1]

4.11.4 References

[1] "Davidson Seamount: In 2009, Monterey Bay National Marine Sanctuary Expanded To Include The Davidson Seamount Management Zone". NOAA (Monterey Bay National Marine Sanctuary). 2009-05-19. Retrieved 2009-11-29.

[2] "Seamount Catalog". *Seamount database*. Earthref.org, a National Science Foundation Project. Retrieved 29 November 2009.

[3] "Geology of Davidson Seamount". NOAA, Office of Ocean Exploration and Research. February 3, 2006. Retrieved December 2, 2009.

[4] "Davidson Seamount" (PDF). NOAA, Monterey Bay National Marine Sanctuary. 2006. Retrieved 2 December 2009.

[5] "Role of Sanctuary in Davidson Seamount Expedition". NOAA, Office of Ocean Exploration and Research. May 19, 2002. Retrieved 2 December 2009.

[6] "Ask an Explorer". NOAA. July 7, 2009. Retrieved December 3, 2009.

[7] "Seamounts may serve as refuges for deep-sea animals that struggle to survive elsewhere". PhysOrg. February 11, 2009. Retrieved December 7, 2009.

[8] McClain, Craig R.; McClain CR., Lundsten L., Ream M., Barry J., DeVogelaere A. (January 7, 2009). Rands, Sean, ed. "Endemicity, Biogeography, Composition, and Community Structure On a Northeast Pacific Seamount". *PLoS ONE* **1** (4): e4141. Bibcode:2009PLoSO...4.4141M. doi:10.1371/journal.pone.0004141. PMC 2613552. PMID 19127302. Retrieved December 3, 2009.

[9] Lundsten, L; L. Lundsten, J. P. Barry, G. M. Cailliet, D. A. Clague, A. DeVogelaere, J. B. Geller (January 13, 2009). "Benthic invertebrate communities on three seamounts off southern and central California". *Marine Ecology Progress Series* (Inter-Research Science Center) **374**: 23–32. doi:10.3354/meps07745.

[10] DeVogelaere, Andrew (July 7, 2009). "Mission Plan". NOAA, Office of Ocean Exploration and Research. Retrieved December 6, 2009.

[11] "Davidson Seamount: Exploring Ancient Coral Gardens". NOAA, Office of Ocean Exploration and Research. December 3, 2009. Retrieved 6 December 2009.

[12] "Ocean expedition to explore the ancient coral gardens on undersea mountain—Public can share discoveries on NOAA website" (PDF). *Press release*. Monterey Bay Aquarium Research Institute (MBARI). January 26, 2006. Retrieved 6 December 2009.

[13] "Sanctuary Science: Davidson Seamount". NOAA. September 18, 2008. Retrieved 8 December 2009.

[14] "NOAA Releases Plans for Managing and Protecting Cordell Bank, Gulf of Farallones and Monterey Bay National Marine Sanctuaries" (PDF). *Press release*. NOAA. November 20, 2008. Retrieved 2 December 2009.

4.11.5 External links

- Deep-sea coral research at Davidson (2006)
- SIMoN images
- SIMon Ecological overview
- Final Rule of incorporation into Monterey Bay

4.12 Dellwood Seamounts

The **Dellwood Seamounts**, also called the **Dellwood Seamount Range** or the **Dellwood Seamount Chain**, is a range of seamounts located in the Pacific Ocean northwest of Vancouver Island, British Columbia, Canada.

4.12.1 References

- British Columbia Marine Topography

4.12.2 See also

- Volcanism of Canada
- Volcanism of Western Canada
- List of volcanoes in Canada

4.13 Denson Seamount

Denson Seamount is a submarine volcano in the Kodiak-Bowie Seamount chain, with an estimated age of 18 million years.[1] It lies at the southern end of the chain near the Canada-United States border. It was one of the underground volcanic extrusions investigated by the 2004 Gulf of Alaska Seamount Expedition.[2] The expedition's goal was:

> "Our goal was to gain an understanding of the geologic histories of the five previously unexplored seamounts in the Gulf of Alaska. To achieve this we created a full-coverage swath bathymetry map of each seamount and its surroundings, and we collected rock samples at all possible depths." *-Randy Keller, Oregon State University*[3]

On August 6, 2004, the DSV Alvin dropped down near the Denson Seamount and collected basaltic rock to try to determine the age of the seamount. The search was difficult because the salty water had altered the volcanic rock over time, but they determined in lab tests that the Denson Seamount was about 18 million years old. The team used bathymetric mapping to render three-dimensional images of the Denson Seamount and its counterparts.

4.13.1 References

[1] NOAA Ocean Explorer: Gulf of Alaska 2004 Retrieved on 2007-09-03

[2] http://oceanexplorer.noaa.gov/explorations/04alaska/welcome.html

[3] http://oceanexplorer.noaa.gov/explorations/04alaska/logs/summary/summary.html

4.13.2 External links

- The expedition's official site

4.14 Eastern Gemini Seamount

The **Eastern Gemini Seamount**, also known as **Oscostar**, is a seamount in the southwestern Pacific Ocean, about halfway between Vanuatu's Tanna and Matthew Islands. The only recorded eruption from Eastern Gemini was observed by a passing ship on February 18, 1996 when bursts of very dark water were observed.*[1]

4.14.1 See also

- List of volcanoes in French Southern and Antarctic Lands

4.14.2 References

[1] "Eastern Gemini Seamount". *Global Volcanism Program*. Smithsonian Institution.

4.15 Explorer Seamount

The **Explorer Seamount** is a seamount located in the Pacific Ocean off the coast of British Columbia, Canada. It is on the Explorer Ridge, a tectonic spreading centre that separates the Pacific and Explorer plates and so the volcanism is rift-related. It is the namesake of the Explorer Ridge.

Explorer Seamount is named after the Coast and Geodetic Survey ship *Explorer*, which ran from 1940 to 1943 in the northern Pacific Ocean and the Gulf of Alaska.*[1]

4.15.1 See also

- Volcanism of Canada

- Volcanism of Western Canada

- List of volcanoes in Canada

4.15.2 Notes

[1] National Geospatial-Intelligence Agency - Undersea Features History

4.15.3 References

• British Columbia Marine Topography

4.16 Ferrel Seamount

Ferrel Seamount is a small seamount (underwater volcano) west of Baja California, at 29°30′52″N 117°17′39″W / 29.51444°N 117.29417°W. Ferrel seamount has been mapped approximately 18% by the USGS, and has two summits. It is located in the Baja California seamounts region, and sits on the edge of the abyssal plain.[*][1]

4.16.1 References

[1] "Detailed Seamount Information - Ferrel Seamount". *Seamounts Database.* Retrieved 2009-03-26.

4.17 Filippo Reef

Filippo Reef is a reef that is asserted to be located in the Pacific Ocean at 05°30′S 151°50′W / 5.500°S 151.833°WCoordinates: 05°30′S 151°50′W / 5.500°S 151.833°W, 450 km east of Starbuck Island in the Line Islands, which was reported by the master of the Italian barque *Filippo* as having been seen on 28 June 1886. From an unidentified report of breakers dated 1926, it was estimated to have a water depth of only 0.6 to 0.9 meters, and appeared to be about 1.6 km long northwest to southeast, and less in width.[*][1]

The only sighting referenced shows a date of 1926. The topographic data supplied by the General Bathymetric Chart of the Oceans (GEBCO), however, shows a sea depth of 3.3 miles.[*][2][*][3] It is therefore likely that this 1926 report is an error, and that Filippo Reef is a phantom island, and that therefore any maps showing it are relying on the 1926 report and are also in error.

Nevertheless, it is marked in the 2005 edition of the *National Geographic Atlas of the World.*[*][4]

4.17.1 References

[1] National Geospatial-Intelligence Agency (2013). "Sector 2. The Line Islands, the Cook Islands, the Samoa Islands, and the Tonga Islands" (PDF). *Pub. 126: Sailing Directions (Enroute): Pacific Islands* (10 ed.). Springfield, Virginia. p. 42. Retrieved 7 May 2013.

[2] General Bathymetric Chart of the Oceans (GEBCO)

[3] "Filippo Reef". *WolframAlpha.com.* Retrieved 7 May 2013.

[4] *National Geographic Atlas of the World* (8th ed.). Washington, D.C.: National Geographic. 2005. p. plate 93. ISBN 0-7922-7543-8.

4.18 Foundation Seamounts

The **Foundation Seamounts**, also called the **Foundation Seamount chain**, are a northwest trending chain of seamounts located in the southern Pacific Ocean between 33° S and 39° S and between 111° W and 125° W (south and southeast of Pitcairn Island). The chain is 1,350 kilometres (839 mi) long and 180 km (112 mi) wide, and is composed by clusters of submarine volcanoes. The age of these submarine volcanoes is unknown.[*][1][*][2]

4.18.1 See also

- *Jasus caveorum*

4.18.2 References

[1] J. Mammerickx (1992). "The Foundation Seamounts – tectonic setting of a newly discovered seamount chain in the South Pacific". *Earth and Planetary Science Letters* **113** (3): 293–306. Bibcode:1992E&PSL.113..293M. doi:10.1016/0012-821X(92)90135-I.

[2] Colin W. Devey, Roger Hékinian, D. Ackermand, N. Binard, B. Francke, C. Hémond, V. Kapsimalis, S. Lorenc, M. Maia, H. Möller, K. Perrot, J. Pracht, T. Rogers, K. Stattegger, S. Steinke & P. Victor (1997). "The Foundation seamount chain: a first survey and sampling". *Marine Geology* **137** (3–4): 191–200. doi:10.1016/S0025-3227(96)00104-1.

4.19 Graham Seamount

The **Graham Seamount** is a seamount located in the Pacific Ocean off the coast of the Queen Charlotte Islands, British Columbia, Canada.

4.19.1 See also

- Volcanology of Canada
- Volcanology of Western Canada
- List of volcanoes in Canada

4.19.2 References

- British Columbia Marine Topography

4.20 Green Seamount

Green Seamount is a small seamount (an underwater volcano) off the western coast of Mexico. It and the nearby Red Seamount were visited in 1982 by an expedition using *DSV Alvin*, which observed the seamount's sedimentary composition, sulfur chimneys, and biology. Thus, Green Seamount is well-characterized for such a small feature.[*][4]

4.20.1 Geology

Green Seamount has a minimal sedimentary cover (0–10 cm (0.0–3.9 in)) of unusually fine sands, with the exception of its more thickly covered caldera (where it was over 50 cm (20 in) thick). Green Seamount is also home to a small number of hydrothermal vents near its caldera wall, which *Alvin* observed (on Red Seamount) to be oxide-rich with temperatures of 13.5 °C (56 °F),[*][4] very low temperatures for a hydrothermal vent.[*][5] Non-hydrothermal sediments on the volcano were observed to be light, "cream-colored" carbonates thinly masked in a finer gray and green sediment.[*][4]

The expedition also noted sulfur mounds near the caldera and pit crater walls;[*][4] these are unusual because, unlike most mid-ocean ridge seamounts, Green's "sulfur chimneys" contain a high amount of silicon, iron, copper, and quartz, but are poor in zinc. The complicated process that created the vents occurred between 140,000 and 70,000 years ago, and by looking at the sulfur chimneys scientists can estimate that it has taken about 260,000 years for the Green Seamount to reach its present height.[*][3] There is also an iron-manganese crust on a small off-axis seamount that adjoins Green Seamount,[*][2] a configuration common with sulfur-rich seamounts.[*][6]

4.20.2 Biology

Green Seamount is not very biologically diverse. The only fauna observed at its hydrothermal sites were single-celled, fan-shaped xenophyophores, and some shrimp. Still, marks of biological life were plentiful wherever sediments were 5 cm (2 in) or more deep.[*][4] The xenophyophores in particular were of a type that was found on many of the seamounts in the eastern Pacific region.[*][7] *Alvin* took two samples, a sponge of the Hexactinellid family and barnacles of the Sessilia order, back to the surface for analysis.[*][8]

4.20.3 References

[1] "Green Seamount". *EarthRef*. National Science Foundation. Retrieved 29 June 2011.

[2] "Green Seamount". *InterRidge Vents Database*. National Oceanography Centre, Southampton. April 10, 2010. Retrieved 29 June 2011.

[3] Jeffery C. Alt et al. (1 January 1987). "Hydrothermal sulfide and oxide deposits on seamounts near 21°N, East Pacific Rise". *Geological Society of America Bulletin* **98** (2): 157. Bibcode:1987GSAB...98..157A. doi:10.1130/0016-7606(1987)98<157:HSAODO>2.0.CO; ISSN 0016-7606. Retrieved 29 June 2011.

[4] Barbara H. Keating et al. (1988). *Seamounts, Islands, and Atolls*. American Geophysical Union. pp. 193–195. ISBN 978-0-87590-068-1. Retrieved 29 June 2011.

[5] Karsten M. Haase et al. (2009). "Fluid compositions and mineralogy of precipitates from Mid Atlantic Ridge hydrothermal vents at 4°48'S". Publishing Network for Geoscientific & Environmental Data (PANGAEA). doi:10.1594/PANGAEA.727454. Retrieved 29 June 2011.

[6] Y. Fouquet et al. (February 15, 1997). "Where are the Large Hydrothermal Sulphide Deposits in the Oceans?". *Philosophical Transactions of the Royal Society A* **355** (1723): 427–441. doi:10.1098/rsta.1997.0015.

[7] Levin, L.A. and C.L. Thomas (1988). "The ecology of xenophyophores (Protista) on eastern Pacific seamounts". *Deep Sea Research Part A. Oceanographic Research Papers* **35** (12): 2003–2027. Bibcode:1988DSRI...35.2003L. doi:10.1016/0198-0149(88)90122-7. Retrieved 29 June 2011.

[8] "Observation Data". *Seamounts Online*. CenSeam. Retrieved 29 June 2011.

4.21 Guide Seamount

Guide Seamount is a seamount in the eastern Pacific Ocean, about 16.6±0.5 million years old. It is similar in shape and orientation to the nearby Davidson, Pioneer, Rodriguez, and Gumdrop seamounts. It is named for the U.S. National Geodetic Survey survey ship the *US&GS Guide*.

Guide Seamount is constructed of four nearly parallel volcanic ridges, separated by sediment-filled throughs. These are aligned parallel to magnetic anomalies in the underlying oceanic crust. It is very similar in shape and structure to the nearby Davidson Seamount, except that it is smaller, at approximately 16.5 km (10 mi) by 5 km (3 mi). It rises about 1,440 m (4,724 ft) above the seafloor and sits at depth of 1,682 m (5,518 ft).[*][1]

The lavas from Guide are mostly alkalic basalt, hawaiite, mugearite with some pyroclastic flows near the top of the summit.[*][1]

4.21.1 References

[1] "Mapping Program: Guide Seamount". MBARI. February 6, 2009. Retrieved 8 December 2009.

[2] "Seamount Catalog". *Seamounts database*. EarthRef, a National Science Foundation project. Retrieved 2009-04-09.

4.21.2 External links

- Interactive bathymetry tool

- Appearance in original 1932 bathymetry

4.22 Gumdrop Seamount

Gumdrop Seamount is a small seamount (underwater volcano) located on the flank of Pioneer Seamount, off the coast of Central California. It is the northernmost of the related seamounts in the region, which includes Davidson, Guide, Pioneer, and Rodriguez seamounts. It is defined by a series of aligned cones separated by troughs filled with sediments, the majority of which are poorly defined. The largest cone rises to within 1,207 m (3,960 ft) of sea level. It is estimated to have a volume of about 100 km^3 (24 cu mi), but the poorly defined base hinders observations of its size. Samples recovered from Gumdrop are highly vesicular in origin, and include alkalic basalt, hawaiite, and mugearite; however, their ages have yet to be determined.[*][2]

4.22.1 References

[1] "Seamount Catalog". *Seamounts database.* EarthRef, a National Science Foundation project. Retrieved 2009-04-09.

[2]

4.23 Heck Seamount

The **Heck Seamount** is a seamount located in the Pacific Ocean off the coast of central Vancouver Island, British Columbia, Canada.

4.23.1 References

- British Columbia Marine Topography

4.23.2 See also

- Volcanism of Canada

- Volcanism of Western Canada

- List of volcanoes in Canada

4.24 Hodgkins Seamount

Hodgkins Seamount is a seamount in the Kodiak-Bowie Seamount chain, located south of Pierce Seamount and north of Bowie Seamount. Hodgkins Seamount has apparently experienced two generically different episodes of volcanism, separated by about 12 million years ago.[*][1] Like the rest of the Kodiak-Bowie seamounts, it was formed by the Bowie hotspot.

4.24.1 See also

- Volcanism of Canada

- Volcanism of Western Canada

- List of volcanoes in Canada

4.24.2 References

[1] Geochronology and origin of the Pratt-Welker seamount chain, Gulf of Alaska: A new pole of rotation for the Pacific plate

4.25 Jasper Seamount

Jasper Seamount is a seamount (underwater volcano) located in the Fieberling-Guadalupe seamount track, west of Baja California, Mexico. Jasper is the site of detailed geophysical geological and geochemical studies which suggest that many seamounts, big and small, follow the same pattern of growth and death that was originally used to describe the Hawaiian - Emperor seamount chain.

Jasper Seamount is an elongated volcano, with a northwest-northeast summit and several volcanic cones on the summit. The base is 4000 meters below sea level and it rises to a peak of 700 meters below sea level. A total of 15 dredge hauls from the seamount have been collected, and ocean-bottom seismometers have been placed to observe earthquake activity. In-depth studies have given scientists a detailed view of the seamount's internal structure.*[1]

The model developed by the Jasper Seamount studies closely resembles that of the Hawaiian islands, especially the eruption, in stages, of increasingly alkalic rocks. More than 90% of the volcano is tholeiitic to alkalic basalt*[1] in a wide variety of forms, including pillow lava and vesicular lapilli, evidence of shallow submarine eruptions in the past, and a wide range of xenoliths.*[2]

Jasper Seamount formed 11.5-10 million years ago. Base volcanics are very similar to those found at Hawaii. The flanks formed about 8.7-7.5 million years ago, and its summit 4.8-4.1 million years ago. The maintenance of a near-constant silicon dioxide ratio in all of the rocks collected is a sign of increasingly lower degrees of mantle melting. Overall Jasper Seamount rock is very similar, even in trace elements, to Hawaiian volcanics, despite Jasper's much smaller size (690 km^3 or 166 cu mi versus ~3,000 km^3 or 720 cu mi). These results seem to show that the Hawaiian model could be applied to other seamounts as well, and that many seamounts go through the same processes regardless of their size.*[1]

4.25.1 See also

- Graveyard Seamounts

- Mud volcano

- Muirfield Seamount

- Sedlo Seamount

- South Chamorro Seamount

4.25.2 References

[1] Jasper G. Konter, Hubert Staudigel, Jeffry Gee. "Spotlight 2: Jasper Seamount" (PDF). *Oceanography*. Seamounts Special Issue (Oceanography Society) **23** (1). Retrieved 13 August 2015.

[2] Jeff Gee, Hubert Staudigel, James H. Natland (1991). "Geology and Petrology of Jasper Seamount". *Journal of Geophysical Research* (American Geophysical Union) **96** (B3): 4083–4105. Bibcode:1991JGR....96.4083G. doi:10.1029/90JB02364. Retrieved 28 July 2010.

4.26   Kavachi

Kavachi is one of the most active submarine volcanoes in the south-west Pacific Ocean.*[2] Located south of Vangunu Island in the Solomon Islands, it is named after a sea god of the New Georgia Group islanders, and is also referred to locally as Rejo te Kavachi ("Kavachi's oven'). The volcano has become emergent and then been eroded back into the sea at least eight times since its first recorded eruption in 1939.*[3]

In May 2000, an international research team aboard the CSIRO research vessel FRANKLIN fixed the position of the volcano at 8° 59.65'S, 157° 58.23'E. At that time the vent of the volcano was below sea level, but frequent eruptions ejected molten lava up to 70m above sea level, and sulfurous steam plumes up to 500m. The team mapped a roughly conical feature rising from 1,100 m water depth, with the volcano having a basal diameter of about 8 km.*[4]*[5]

When the volcano erupted in 2003, a 15-meter-high island formed above the surface, but it disappeared soon after. Additional eruptive activity was observed and reported in March 2004 and April 2007.*[6]*[7]

Recently, marine wildlife has been found living inside Kavachi including two species of sharks and a sixgill stingray. Scientists are baffled and excited about this phenomenon.*[8]

4.26.1   See also

- List of recently born islands

- List of volcanoes in the Solomon Islands

- Woodlark Basin

4.26.2   References

[1]   "Kavachi" . *Global Volcanism Program*. Smithsonian Institution. Retrieved 2010-02-26.

[2]   "Kavachi" . *Global Volcanism Program*. Smithsonian Institution.

[3]   "Kavachi - Eruptive History" . *Global Volcanism Program*. Smithsonian Institution.

[4]   "Fiery birth of new Pacific Island" . Commonwealth Scientific and Industrial Research Organisation (CSIRO). 24 May 2000.

[5]   "Volcano Island Born" . *All Things Considered*. US National Public Radio. 26 May 2000.

[6]   "Kavachi - Monthly Reports" . *Global Volcanism Program*. Smithsonian Institution.

[7]   "Kavachi Submarine Volcano" . Corey Howell, The Wilderness Lodge.

[8]   "Sharks Discovered Inside Underwater Volcano" .

4.26.3   Further reading

- Baker, E.T., Massoth, G.J., de Ronde, C.E.J., Lupton, J.E., Lebon, G., and McInnes, B.I.A. 2002. Observations and sampling of an ongoing subsurface eruption of Kavachi volcano, Solomon Islands, May 2000, Geology, 30 (11), 975-978. (geology.geoscienceworld.org/cgi/reprint/30/11/975.pdf)

- Dunkley, P.M., 1983. Volcanism and the evolution of the ensimatic Solomon Islands Arc, in Shimozuro, D. And Yokoyama, I.,(eds.), Arc Volcanism: Physics and Tectonics. Tokyo, Terrapub, 225-241.

- Johnson, R.W. and Tuni, D. 1987. Kavachi, an active forearc volcano in the western Solomon Islands: Reported eruptions between 1950 and 1982, in B. Taylor and N.F. Exon, (eds.), 1987, Marine Geology, Geophysics, and Geochemistry of the Woodlark Basin-Solomon Islands, Circum-Pacific Council for Energy and Mineral Resources Earth Science Series, v. 7: Houston, Texas, Circum-Pacific Council for Energy and Mineral Resources.

4.26.4 External links

- Kavachi in action

4.27 Lō'ihi Seamount

Lō'ihi Seamount is an active submarine volcano located around 35 km (22 mi) off the southeast coast of the island of Hawai'i[6] about 975 m (3,000 ft) below sea level. This seamount lies on the flank of Mauna Loa, the largest shield volcano on Earth. Lō'ihi meaning "long" in Hawaiian, is the newest volcano in the Hawaiian-Emperor seamount chain, a string of volcanoes that stretches over 5,800 km (3,600 mi) northwest of Lō'ihi. Unlike most active volcanoes in the Pacific Ocean that make up the active plate margins on the Pacific Ring of Fire, Lō'ihi and the other volcanoes of the Hawaiian-Emperor seamount chain are hotspot volcanoes and formed well away from the nearest plate boundary. Volcanoes in the Hawaiian Islands arise from the Hawai'i hotspot, and as the youngest volcano in the chain, Lō'ihi is the only Hawaiian volcano in the deep submarine preshield stage of development.

Lō'ihi began forming around 400,000 years ago and is expected to begin emerging above sea level about 10,000–100,000 years from now. At its summit, Lō'ihi Seamount stands more than 3,000 m (10,000 ft) above the seafloor, making it taller than Mount St. Helens was before its catastrophic 1980 eruption. The summit is currently 975 m (3,000 ft) below sea level. A diverse microbial community resides around Lō'ihi's many hydrothermal vents.

In the summer of 1996, a swarm of 4,070 earthquakes was recorded at Lō'ihi. This series included more earthquakes than any other swarm in Hawaiian recorded history. The swarm altered 10 to 13 square kilometres (4 to 5 sq mi) of the seamount's summit; one section, *Pele's Vents*, collapsed entirely upon itself and formed the renamed *Pele's Pit*. The volcano has remained relatively active since the 1996 swarm and is monitored by the National Oceanic and Atmospheric Administration (NOAA) and the United States Geological Survey (USGS). The Hawaii Undersea Geological Observatory (HUGO) provided real-time data on Lō'ihi between 1997 and 2002. Lō'ihi last erupted in 1996, before the earthquake swarm of that summer.

4.27.1 Characteristics

Geology

See also: Evolution of Hawaiian volcanoes

Lō'ihi is a seamount, or underwater volcano, on the flank of Mauna Loa, the Earth's largest shield volcano. It is the newest volcano created by the Hawai'i hotspot in the extensive Hawaiian-Emperor seamount chain. The distance between the summit of the older Mauna Loa and the summit of Lō'ihi is about 80 km (50 mi), which is, coincidentally, also the approximate diameter of the Hawai'i hotspot.[4] Lō'ihi consists of a summit area with three pit craters, an 11 km (7 mi) long rift zone extending north from the summit, and a 19 km (12 mi) long rift zone extending south-southeast from the summit.[7]

The summit's pit craters are named West Pit, East Pit, and Pele's Pit.[8] Pele's Pit is the youngest of this group and is located at the southern part of the summit. The walls of Pele's Pit stand 200 m (700 ft) high and were formed in July 1996 when its predecessor, Pele's Vent, a hydrothermal field near Lō'ihi's summit, collapsed into a large depression.[6] The thick crater walls of Pele's Pit—averaging 20 m (70 ft) in width, unusually thick for Hawaiian volcanic craters—suggest its craters have filled with lava multiple times in the past.[9]

Lō'ihi's north–south trending rift zones create a distinctive elongated shape, from which the volcano's Hawaiian name, meaning "long", derives.[10] The north rift zone consists of a longer western portion and a shorter eastern rift zone. Observations show that both the north and south rift zones lack sediment cover, indicating recent activity. A bulge in the western part of the north rift zone contains three 60–80 m (200–260 ft) cone-shaped prominences.[9]

Until 1970, Lō'ihi was thought to be an inactive volcano that had been transported to its current location by sea-floor spreading. The seafloor under Hawai'i is 80–100 million years old and was created at the East Pacific Rise, an oceanic spreading center where new sea floor forms from magma that erupts from the mantle. New oceanic crust moves away from the spreading center. Over a period of 80–100 million years, the sea floor under Hawai'i moved from the East Pacific Rise to its present location 6,000 km (4,000 mi) west, carrying ancient seamounts with it. When scientists investigated a series of earthquakes off Hawai'i in 1970, they discovered that Lō'ihi was an active member of the Hawaiian-Emperor seamount chain.

Lō'ihi is built on the seafloor with a slope of about five degrees. Its northern base on the flank of Mauna Loa is 1,900 m (6,200 ft) below sea level, but its southern base is a more substantial 4,755 m (15,600 ft) below the surface. Thus, the summit is 931 m (3,054 ft) above the seafloor as measured from the base of its north flank, but 3,786 m (12,421 ft) high when measured from the base of its southern flank.[4]

Lō'ihi is following the pattern of development that is characteristic of all Hawai'ian volcanoes. Geochemical evidence from Lō'ihi lavas indicates that Lō'ihi is in transition between the preshield and shield volcano stage, providing valuable clues to the early development of Hawaiian volcanoes. In the preshield stage, Hawaiian volcanoes have steeper sides and a lower level of activity, producing an alkali basalt lava.[11][12] Continued volcanism is expected to eventually create an island at Lō'ihi. Lō'ihi experiences frequent landslides; the growth of the volcano has destabilized its slopes, and extensive areas of debris inhabit the steep southeastern face. Similar deposits from other Hawaiian volcanoes indicate that landslide debris is an important product of Hawaiian volcanos' early development.[5] Lō'ihi is predicted to rise above the surface in 10,000 to 100,000 years.[1]

Age and growth

Radiometric dating was used to determine the age of rock samples from Lō'ihi. The Hawaii Center for Volcanology tested samples recovered by various expeditions, notably the 1978 expedition, which provided 17 dredge samples. Most of the samples were found to be of ancient origin; the oldest dated rock is around 300,000 years old. Following the 1996 event, some young breccia was also collected. Based on the samples, scientists estimate Lō'ihi is about 400,000 years old. The rock accumulates at an average rate of 3.5 mm (0.14 in) per year near the base, and 7.8 mm (0.31 in) near the summit. If the data model from other volcanoes such as Kīlauea holds true for Lō'ihi, 40% of the volcano's mass formed within the last 100,000 years. Assuming a linear growth rate, Lō'ihi is 250,000 years old. However, as with all hotspot volcanoes, Lō'ihi's level of activity has increased with time; therefore, it would take at least 400,000 years for such a volcano to reach Lō'ihi's mass.[5] As Hawaiian volcanoes drift northwest at a rate of about 10 cm (4 in) a year, Lō'ihi was 40 km (25 mi) southeast of its current position at the time of its initial eruption.[13]

4.27.2   Activity

Lō'ihi is a young and fairly active volcano, although less active than nearby Kīlauea. In the past few decades, several earthquake swarms have been attributed to Lō'ihi, the largest of which are summarized in the table below.[14] The volcano's activity is now known to predate scientific record keeping of its activity, which commenced in 1959.[15] Most earthquake swarms at Lō'ihi have lasted less than two days; the two exceptions are the 1991–92 earthquake, lasting several months, and the 1996 event, which was shorter but much more pronounced. Both of the earthquakes followed a pattern of activity that began on the flank and migrated to the summit. The 1996 event was directly observed by an autonomous ocean bottom observatory (OBO), allowing scientists to calculate the depth of the earthquakes as 6 km (4 mi) to 8 km (5 mi) below the summit, approximating to the position of Lō'ihi's extremely shallow magma chamber.[5] This is evidence that Lō'ihi's seismicity is volcanic in origin.[8]

The low level seismic activity documented on Lō'ihi since 1959 has shown that between two and ten earthquakes per month are traceable to the summit.[15] Earthquake swarm data have been used to analyze how well Lō'ihi's rocks propagate seismic waves and to investigate the relationship between earthquakes and eruptions. This low level activity is periodically punctuated by large swarms of earthquakes, each swarm comprising up to hundreds of earthquakes. The majority of the earthquakes are not distributed close to the summit, though they follow a north–south trend. Rather, most of the earthquakes occur in the southwest portion of Lō'ihi.[5] The largest recorded swarms took place on Lō'ihi in 1971, 1972, 1975, 1991–92 and 1996. The nearest seismic station is around 30 km (20 mi) from Lō'ihi, on the south coast of Hawai'i. Seismic events that have a magnitude under 2 are recorded often, but their location cannot be determined precisely as it can for larger events.[16] In fact, HUGO (Hawai'i Undersea Geological Observatory), positioned on Lō'ihi's flank, detected ten times as many earthquakes as were recorded by the Hawaiian Volcano Observatory (HVO) seismic network.[*][5]

1996 earthquake swarm

The largest amount of activity recorded for the Lōʻihi seamount was a swarm of 4,070 earthquakes between July 16 and August 9, 1996.[*][4] This series of earthquakes was the largest recorded for any Hawaiian volcano to date in both amount and intensity. Most of the earthquakes had moment magnitudes of less than 3.0. "Several hundred" had a magnitude greater than 3.0, including more than 40 greater than 4.0 and a 5.0 tremor.[*][16][*][18]

The final two weeks of the earthquake swarm were observed by a quick response cruise launched in August 1996. The National Science Foundation funded an expedition by University of Hawaiʻi scientists, led by Frederick Duennebier, that began investigating the swarm and its origin in August 1996. The scientists' assessment laid the groundwork for many of the expeditions that followed.[*][19] Follow-up expeditions to Lōʻihi took place, including a series of manned-submersible dives in August and September. These were supplemented by a great deal of shore-based research.[*][18] Fresh rock collected during the expedition revealed that an eruption occurred *before* the earthquake swarm.[*][20]

Submersible dives in August were followed by NOAA-funded research in September and October 1996. These more detailed studies showed the southern portion of Lōʻihi's summit had collapsed, a result of a swarm of earthquakes and the rapid withdrawal of magma from the volcano. A crater 1 km (0.6 mi) across and 300 m (1,000 ft) deep formed out of the rubble. The event involved the movement of 100 million cubic meters of volcanic material. A region of 10 to 13 km^2 (4 to 5 sq mi) of the summit was altered and populated by bus-sized pillow lava blocks, precariously perched along the outer rim of the newly formed crater. "Pele's Vents," an area on the southern side, previously considered stable, collapsed completely into a giant pit, renamed "Pele's Pit". Strong currents make submersible diving hazardous in the region.[*][19]

The researchers were continually met by clouds of sulfide and sulfate. The sudden collapse of Pele's Vents caused a large discharge of hydrothermal material. The presence of certain indicator minerals in the mixture suggested temperatures exceeded 250 °C, a record for an underwater volcano. The composition of the materials was similar to that of black smokers, the hydrothermal vent plumes located along mid-ocean ridges. Samples from mounds built by discharges from the hydrothermal plumes resembled white smokers.[*][21]

The studies demonstrated that the most volcanically and hydrothermally active area was along the southern rift. Dives on the less active northern rim indicated that the terrain was more stable there, and high lava columns were still standing upright.[*][19] A new hydrothermal vent field (Naha Vents) was located in the upper-south rift zone, at a depth of 1,325 m (4,350 ft).[*][5][*][22]

Recent activity

Lōʻihi has remained largely quiet since the 1996 event; no activity was recorded from 2002 to 2004. The seamount showed signs of life again in 2005 by generating an earthquake bigger than any previously recorded there. USGS-ANSS (Advanced National Seismic System) reported two earthquakes, magnitudes 5.1 and 5.4, on May 13 and July 17. Both originated from a depth of 44 km (27 mi). On April 23, a magnitude 4.3 earthquake was recorded at a depth of approximately 33 km (21 mi). Between December 7, 2005, and January 18, 2006, a swarm of around 100 earthquakes occurred, the largest measuring 4 on the Moment magnitude scale and 12 km (7 mi) to 28 km (17 mi) deep. Another earthquake measuring 4.7 was later recorded approximately midway between Lōʻihi and Pāhala (on the south coast of the island of Hawaiʻi).[*][14]

4.27.3 Exploration

Early work

Expedition Timeline

Lōʻihi Seamount's first depiction on a map was on Survey Chart 4115, a bathymetric rendering of part of Hawaiʻi compiled by the US Coast and Geodetic Survey in 1940. At the time, the seamount was non-notable, being one of many in the region. A large earthquake swarm first brought attention to it in 1952. That same year, geologist Gordon A. MacDonald hypothesized that the seamount was actually an active submarine shield volcano, similar to the two active Hawaiian volcanoes, Mauna Loa and Kīlauea. Macdonald's hypothesis placed the seamount as the newest volcano in the Hawaiian-Emperor seamount chain, created by the Hawaiʻi hotspot. However, because the earthquakes were oriented east–west (the direction of the volcanic fault) and there was no volcanic tremor in seismometers distant from the seamount, Macdonald attributed the earthquake to faulting rather than a volcanic eruption.*[5]

Geologists suspected the seamount could be an active undersea volcano, but without evidence the idea remained speculative. The volcano was largely ignored after the 1952 event, and was often mislabeled as an "older volcanic feature" in subsequent charts.*[5] Geologist Kenneth O. Emery is credited with naming the seamount in 1955,*[7] describing the long and narrow shape of the volcano as *Lōʻihi*, the Hawaiian word for "long".*[10]*[26] In 1978, an expedition studied intense, repeated seismic activity known as earthquake swarms in and around the Lōʻihi area. Rather than finding an old, extinct seamount, data collected revealed Lōʻihi to be a young, possibly active volcano. Observations showed volcano to be encrusted with young and old lava flows as well as actively venting hydrothermal fluids.*[2]

In 1978, a US Geological Survey research ship collected dredge samples and photographed Lōʻihi's summit with the goal of studying whether Lōʻihi is active. Analysis of the photos and testing of pillow lava rock samples appeared to show that the material was "fresh", yielding more evidence that Lōʻihi is still active. An expedition from October 1980 to January 1981 collected further dredge samples and photographs, providing additional confirmation.*[27] Studies indicated that the eruptions came from the southern part of the rift crater. This area is closest to the Hawaiʻi hotspot, which supplies Lōʻihi with magma.*[5] Following a 1986 seismic event, a network of five ocean bottom observatories (OBOs) were deployed on Lōʻihi for a month. Lōʻihi's frequent seismicity makes it an ideal candidate for seismic study through OBOs.*[5] In 1987, the submersible DSV Alvin was used to survey Lōʻihi.*[28] Another autonomous observatory was positioned on Lōʻihi in 1991 to track earthquake swarms.*[5]

1996 to present

The bulk of information about Lōʻihi comes from dives made in response to the 1996 eruption. In a dive conducted almost immediately after seismic activity was reported, visibility was greatly reduced by high concentrations of displaced minerals and large floating mats of bacteria in the water. The bacteria that feed on the dissolved nutrients had already begun colonizing the new hydrothermal vents at Pele's Pit (formed from the collapse of the old ones), and may be indicators of the kinds of material ejected from the newly formed vents. They were carefully sampled for further analysis in a laboratory.*[19] An OBO briefly sat on the summit before a more permanent probe could be installed.*[29]

Repeated multibeam bathymetric mapping was used to measure the changes in the summit following the 1996 collapse. Hydrothermal plume surveys confirmed changes in the energy, and dissolved minerals emanating from Lōʻihi. Hawaiʻi Undersea Research Laboratory, HURL's 2,000 m (6,562 ft) submersible *Pisces V* allowed scientists to sample the vent waters, microorganisms and hydrothermal mineral deposits.*[6]

In 1997, scientists from the University of Hawaiʻi installed an ocean bottom observatory (OBO) on the summit of Lōʻihi Seamount.*[14] The submarine observatory was nicknamed HUGO, (Hawaiʻi Undersea Geological Observatory). HUGO was connected to the shore, 34 km (21 mi) away, by a fiber optic cable. It gave scientists real-time seismic, chemical and visual data about the state of Lōʻihi, which had by then become an international laboratory for the study of undersea volcanism.*[19] The cable that provided HUGO with power and communications broke in October 1998, effectively shutting it down. On January 19 of the following year, HUGO was visited by *Pisces V*. The observatory functioned for four years before it went dead again in 2002.*[30]

Since 2006, the Fe-Oxidizing Microbial Observatory (FeMO), funded by the National Science Foundation and Microbial Observatory Program, has led cruises to Lōʻihi investigate its microbiology every October. The first cruise, on the ship *R/V Melville* and exploiting the submersible *JASON2*, lasted from September 22 to October 9. These cruises study the large number of Fe-oxidizing bacteria that have colonized Lōʻihi. Lōʻihi's extensive vent system is characterized by a high concentration of CO_2 and Iron, while being low in sulfide. These characteristics make a perfect environment for iron-oxidizing bacteria, called FeOB, to thrive in.*[25]

4.27.4 Ecology

Hydrothermal vent geochemistry

Lōʻihi's mid-Pacific location and its well-sustained hydrothermal system contribute to a rich oasis for a microbial ecosystem. Areas of extensive hydrothermal venting are found on Lōʻihi's crater floor and north slope,[6] and along the summit of Lōʻihi itself. Active hydrothermal vents were first discovered at Lōʻihi in the late 1980s. These vents are remarkably similar to those found at the mid-ocean ridges, with similar composition and thermal differences. The two most prominent vent fields are at the summit: *Pele's Pit* (formally *Pele's Vents*) and *Kapo's Vents*. They are named after the Hawaiian deity Pele and her sister Kapo. These vents were considered "low temperature vents" because their waters were only about 30 °C. The volcanic eruption of 1996 and the creation of Pele's Pit changed this, and initiated high temperature venting; exit temperatures were measured at 77 °C in 1996.[22]

Microorganisms

The vents lie 1,100 m (3,600 ft) to 1,325 m (4,347 ft) below the surface, and range in temperature from 10 to over 200 °C.[22][31] The vent fluids are characterized by a high concentration of CO2 (up to 17 mM) and Fe (Iron), but low in sulfide. Low oxygen and pH levels are important factors in supporting the high amounts of Fe (iron), one of the hallmark features of Lōʻihi. These characteristics make a perfect environment for iron-oxidizing bacteria, called FeOB, to thrive in.[25] An example of these species is *Mariprofundus ferrooxydans*, sole member of the class Zetaproteobacteria.[32] The composition of the materials was similar to that of black smokers, that are a habitat of archaea extremophiles. Dissolution and oxidation of the mineral observed over the next two years suggests the sulfate is not easily preserved.[21]

A diverse community of microbial mats surround the vents and virtually cover Pele's Pit. The Hawaiʻi Undersea Research Laboratory (HURL), NOAA's Research Center for Hawaiʻi and the Western Pacific, monitors and researches the hydrothermal systems and studies the local community.[6] The National Science Foundation (NSF) funded an extremophile sampling expedition to Lōʻihi in 1999. Microbial mats surrounded the 160 °C vents, and included a novel jelly-like organism. Samples were collected for study at NSF's Marine Bioproducts Engineering Center (MarBEC).[6] In 2001, *Pisces V* collected samples of the organisms and brought them to the surface for study.[19]

NOAA's National Undersea Research Center and NSF's Marine Bioproducts Engineering Center are cooperating to sample and research the local bacteria and archaea extremophiles.[6] The fourth FeMO (Fe-Oxidizing Microbial Observatory) cruise occurred during October 2009.[33]

Macroorganisms

Marine life inhabiting the waters around Lōʻihi is not as diverse as life at other, less active seamounts. Fish found living near Lōʻihi include the Celebes monkfish (*Sladenia remiger*), and members of the Cutthroat eel family, Synaphobranchidae.[34] Invertebrates identified in the area include two species endemic to the hydrothermal vents, a bresiliid shrimp (*Opaepele loihi*) of the family Alvinocarididae (described in 1995), and a tube or pogonophoran worm. Dives conducted after the 1996 earthquake swarms were unable to find either the shrimp or the worm, and it is not known if there are lasting effects on these species.[35]

From 1982 to 1992, researchers in Hawaiʻi Undersea Research Laboratory submersibles photographed the fish of Lōʻihi Seamount, Johnston Atoll, and Cross Seamount at depths between 40 m (130 ft) and 2,000 m (6,600 ft).[36][37] A small number of species identified at Lōʻihi were newly recorded sightings in Hawaiʻi, including the Tassled coffinfish (*Chaunax fimbriatus*), and the Celebes monkfish.[36]

4.27.5 See also

- List of volcanoes in the Hawaiian – Emperor seamount chain

4.27.6 References

[1] "Loihi". *Global Volcanism Program*. Smithsonian Institution. Retrieved 2009-03-01.

[2] Rubin, Ken (2006-01-19). "General Information About Loihi". *Hawaii Center for Volcanology*. SOEST. Retrieved 2009-02-01.

[3] Pukui, M.K. & Elbert, S.H. *Hawaiian Dictionary* (1986) University of Hawaii Press ISBN 0-8248-0703-0

[4] "Lōʻihi Seamount Hawaiʻi's Youngest Submarine Volcano". *Hawaiian Volcano Observatory*. United States Geological Survey. Retrieved 2009-03-01.

[5] Michael O. Garcia, Jackie Caplan-Auerbach, Eric H. De Carlo, M.D. Kurz, N. Becker (2005). "Geochemistry, and Earthquake History of Lōʻihi Seamount, Hawaiʻi's youngest volcano" (PDF). *Chemie der Erde – Geochemistry* **66** (2): 81–108. doi:10.1016/j.chemer.2005.09.002. Retrieved 2009-03-20.

[6] Malahoff, Alexander (2000-12-18). "Lōʻihi Submarine Volcano: A unique, natural extremophile laboratory". Office of Oceanic and Atmospheric Research (NOAA). Retrieved 2009-03-01.

[7] Malahoff, Alexander (1987). "Geology of the summit of Lōʻihi submarine volcano". In Decker, Robert W. Wright, Thomas L. Stauffer, Peter H. *Volcanism in Hawaiʻi: U.S. Geological Survey Professional Paper 1350*. United States Geological Survey Professional Paper 1350 **1**. Washington: United States Government Printing Office. pp. 133–44. Retrieved 2009-06-15.

[8] Malahoff, Alexander; Kolotyrkina, Irina Ya.; Midson, Brian P.; Massoth, Gary J. (2006-01-06). "A decade of exploring a submarine intraplate volcano: Hydrothermal manganese and iron at Lōʻihi volcano, Hawaiʻi" (PDF). *G³: Geochemistry, Geophysics, Geosystems* (American Geophysical Union and the Geochemical Society) **7** (6): Q06002. Bibcode:2006GGG.....706002M. doi:10.1029/2005GC001222. ISSN 1525-2027. Retrieved 2009-06-15.

[9] Fornari, D.J., Garcia, M.O., Tyce, R.C., Gallo, D.G. (1988). "Morphology and structure of Loihi seamount based on seabeam sonar mapping". *Journal of Geophysics Research* **93** (15): 227–38. doi:10.1029/jb093ib12p15227. Archived from the original on 2009-04-16. Retrieved 2009-06-14.

[10] *Lōʻihi*, meaning "length, height, distance; long". See: Pukui, Mary Kawena; Samuel Hoyt Elbert (1986). *Hawaiian dictionary: Hawaiian-English, English-Hawaiian*. University of Hawaiʻi Press. p. 209. ISBN 0-8248-0703-0.

[11] Best, Myron G. (1991). *Igneous and Metamorphic Petrology*. Wiley, John & Sons, Incorporated. p. 359. ISBN 978-1-4051-0588-0.

[12] "Evolution of Hawaiian Volcanoes". *Hawaiian Volcano Observatory*. USGS. September 8, 1995. Retrieved 2009-03-07.

[13] Garcia, M.O., Grooms, D., Naughton, J. (1987). "Petrology and geochronology of volcanic rocks". *Lithosphere* (The Geological Society of America) (20): 323–36.

[14] Rubin, Ken (2006-01-20). "Recent Activity at Loihi Volcano – Updates on Geologic Activity at Loihi". *Hawaii Center For Volcanology*. SOEST. Retrieved 2009-03-07.

[15] Caplan-Auerbach, Jackie (1998-07-22). "Recent Seismicity at Loihi Volcano". *Hawaii Center for Volcanology*. SOEST. Retrieved 2009-03-15.

[16] "Loihi – Monthly Reports". *Global Volcanism Program*. Smithsonian Institution. Retrieved 2009-03-13.

[17] "Loihi – Eruptive History". *Global Volcanism Program*. Smithsonian Institution. Retrieved 2009-03-13. Dates for older eruptions retrieved through Isotope dating.

[18] Rubin, Ken (1998-07-22). "The 1996 Eruption and July–August Seismic Event". *Hawaii Center for Volcanology*. SOEST. Retrieved 2009-03-01.

[19] "HURL Current Research: Loihi after the July–August event". *1999 Research*. SOEST. 2001. Archived from the original on 2009-03-05. Retrieved 2009-03-01.

[20] Garcia, M.O., Graham, D.W., Muenow, D.W., Spencer, K., Rubin, K.H., Norman, M.D. (1998). "Petrology and geochronology of basalt breccia from the 1996 earthquake swarm of Loihi seamount, Hawaii: magmatic history of its 1996 eruption". *Bulletin of Volcanology* **59** (8): 577–92. Bibcode:1998BVol...59..577G. doi:10.1007/s004450050211. ISSN 0258-8900. Retrieved 2009-06-13.

[21] Davis, Alicé S.; David A. Clague; Robert A. Zierenberg; C. Geoffrey Wheat; Brian L. Cousens (Apr 2003). "Sulfide formation related to changes in the hydrothermal system on Lōʻihi Seamount, Hawaiʻi, following the seismic event in 1996". *The Canadian Mineralogist* (Mineralogical Association of Canada) **41** (2): 57–472. doi:10.2113/gscanmin.41.2.457.

[22] Rubin, Ken (1998-07-22). "Recent Activity at Loihi Volcano: Hydrothermal Vent and Buoyant Plume Studies". *Hawaii Center for Volcanology*. SOEST. Retrieved 2009-03-15.

[23] Rubin, Ken. "Cruises to Lōʻihi Since the 1996 Eruption and Seismic Swarm". *Hawaiʻi Center for Volcanology*. SOEST. Retrieved 2009-03-15.

[24] Duennebier, Fred (2002-10-01). "HUGO: Update and Current Status". SOEST. Retrieved 2009-03-17.

[25] "Introduction to the Biology and Geology of Lōʻihi Seamount". *Lōʻihi Seamount*. Fe-Oxidizing Microbial Observatory (FeMO). 2009-02-01. Retrieved 2009-03-02.

[26] Cooke, Sarah (Apr 2002). "Lōʻihi and the Hawaiian Hot Spot". *Caltech Undergraduate Research Journal* (California Institute of Technology) **2** (1). Archived from the original on 2005-04-26.

[27] Macdonald, Gordon A.; Agatin T. Abbott; Frank L. Peterson (1983) [1970]. *Volcanoes in the Sea: The Geology of Hawaiʻi* (2nd ed.). Honolulu: University of Hawaiʻi Press. ISBN 0-8248-0832-0.

[28] Garcia, M.O., Irving, A.J., Jorgenson, B.A., Mahoney, J.J., Ito, E. (1993). "An evaluation of temporal geochemical evolution of Loihi summit lavas: results from Alvin submersible dives". *Journal of Geophysical Research* **98** (B1): 537–550. Bibcode:1993JGR....98..537G. doi:10.1029/92JB01707. Retrieved 2009-06-13.

[29] Bryan, Carol; Cooper, P. (December 1995). "Ocean-bottom seismometer observations of seismic activity at Loihi". *Marine Geophysical Researches* (Springer Netherlands) **17**: 485–501. Bibcode:1995MarGR..17..485B. doi:10.1007/BF01204340. ISSN 0025-3235. Retrieved 2009-06-13.

[30] "HUGO: The Hawaiʻi Undersea Geo-Observatory". SOEST. Retrieved 2009-03-15.

[31] Emerson, David; Craig L. Moyer (June 2002). "Neutrophilic Fe-Oxidizing Bacteria Are Abundant at the Lōʻihi Seamount Hydrothermal Vents and Play a Major Role in Fe Oxide Deposition". *Applied and Environmental Microbiology* (American Society for Microbiology) **68** (6): 3085–93. doi:10.1128/AEM.68.6.3085-3093.2002. PMC 123976. PMID 12039770. Retrieved 2009-03-15.

[32] Emerson, David; Rentz, Jeremy A.; Lilburn, Timothy G.; Davis, Richard E.; Aldrich, Henry; Chan, Clara; Moyer, Craig L. (2007). Reysenbach, Anna-Louise, ed. "A novel lineage of proteobacteria involved in formation of marine Fe-oxidizing microbial mat communities". *PLoS ONE* **2** (8): e667. Bibcode:2007PLoSO...2..667E. doi:10.1371/journal.pone.0000667. PMC 1930151. PMID 17668050.

[33] "FeMO4 Dive Cruise 2009". *FeMO*. EarthRef.org. 2009-10-17. Retrieved 2010-02-08.

[34] Rubin, Ken (1998-09-07). "A Tour of Loihi". *Hawaii Center for Volcanology*. SOEST. Retrieved 2009-03-15.

[35] Rubin, Ken (1998-07-22). "Recent Activity at Loihi Volcano – 1996 Seismic/Volcanic Event Summary". *Hawaii Center For Volcanology*. SOEST. Retrieved 2009-05-30. The only two vent-specific macrofaunal species described from Loihi have been a novel bresiliid shrimp, Opaepele loihi (Williams and Dobbs, 1995), and a unique lineage of pogonophoran worm (R. Vrijenhoek, pers. comm.). The post-event dives, however, found no evidence for either, and the long-term impact of the event on these species is unknown.

[36] Chave, E.H.; B.C. Mundy (1994). "Deep-Sea Benthic Fish of the Hawaiian Archipelago, Cross Seamount, and Johnston Atoll". *Pacific Science* (University of Hawaiʻi) **48** (4): 367–409. Retrieved 2009-03-16.

[37] Data from Chave, E.H. and B.C. Mundy (1994) and Scripps Institution of Oceanography (2002). "Observation Data". University of the Pacific. Retrieved 2009-03-16.

4.27.7 Further reading

- Auerbach-Caplan, J.; F. Duennebier (2001-05-25). "Seismic and acoustic signals detected at Loʻihi Seamount by the Hawaiʻi Undersea Geo-Observatory" (PDF). *Geochemistry Geophysics Geosystems* (American Geophysical Union and the Geochemical Society) **2** (5): 1525–2027. Bibcode:2001GGG.....2.1024C. doi:10.1029/2000GC000113. ISSN 1525-2027. Retrieved 2009-04-27.

- Chave, E. H.; Alexander Malahoff (1998). *In Deeper Waters: Photographic Studies of Hawaiian Deep-sea Habitats and Life-forms.* University of Hawaiʻi Press. ISBN 0-8248-2003-7.

- F. K. Duennebier, N. C. Becker, J. Caplan-Auerbach, D. A. Clague, J. Cowen, M. Cremer, M. Garcia, F. Goff, A. Malahoff, G. M. McMurtry, B. P. Midson, C. L. Moyer, M. Norman, P. Okubo, J. A. Resing, J. M. Rhodes, K. Rubin, F. J. Sansone, J. R. Smith, K. Spencer, X. Wen, and C. G. Wheat (1997-06-03). "Researchers rapidly respond to submarine activity at Loʻihi volcano, Hawaiʻi" (PDF). *Eos, Transactions, American Geophysical Union* **78** (22): 229–33. Bibcode:1997EOSTr..78Q.229T. doi:10.1029/97EO00150.

- Emery, K.O. (1955). "Submarine topography south of Hawaiʻi". *Pacific Science* (University of Hawaiʻi Press) **9**: 286–91.

- "Loihi – Data Sources". *Global Volcanism Program.* Smithsonian Institution.

- Klein, F. (1982). "Earthquakes at Loʻihi submarine volcano and the Hawaiian hot spot". *Journal of Geophysical Research* **87**: B9. Bibcode:1982JGR....87....9K. doi:10.1029/JA087iA01p00009. ISSN 0148-0227.

- Macdonald, G.A. (1952). "The South Hawaiʻi Earthquakes of March and April, 1952." *The Volcano Letter.* U.S. Geological Survey Professional Paper 515.

- Malahoff, Alexander; Gary M. McMurtry; John C. Wiltshire; Hsueh-Wen Yeh (1982-07-15). "Geology and chemistry of hydrothermal deposits from active submarine volcano Loʻihi, Hawaiʻi". *Nature* (Nature Publishing Group) **298** (5871): 234–39. Bibcode:1982Natur.298..234M. doi:10.1038/298234a0.

- Malahoff, A.; Gregory, T.; Bossuyt, A.; Donachie, S.; Alarn, M. (Oct 2002). "A seamless system for the collection and cultivation of extremophiles from deep-ocean hydrothermal vents". *IEEE Journal of Oceanic Engineering* (IEEE Oceanic Engineering Society) **27** (4): 862–69. doi:10.1109/JOE.2002.804058.

- J.G. Moore, D.A. Clague, W.R. Normark (Feb 1982). "Diverse basalt types from Loʻihi Seamount, Hawaiʻi". *Geology* (Geological Society of America) **10** (2): 88–92. Bibcode:1982Geo....10...88M. doi:10.1130/0091-7613(1982)10<88:DBTFLS>2.0.CO;2.

- Scripps Institution of Oceanography. (2002). Benthic Invertebrate Collection Database.

4.27.8 External links

- Hawaii Center for Volcanology, University of Hawaiʻi.

- Loʻihi Seamount—USGS website.

- Loihi Submarine Volcano: A unique, natural extremophile laboratory—NOAA research site.

- HURL Current Research – Loihi after the July–August event, on the 1996 Loʻihi Seamount Exploration

- Recent volcanic activity at Loihi – University of Hawaiʻi

- Fe-Oxidizing Microbial Observatory Project (FeMO) Webpage – Earthref.org

4.28 Marisla Seamount

Marisla Seamount, also known as "El Bajo", is located about 8 miles (13 km) north-northeast of La Paz, Mexico. There are three underwater peaks arrayed three hundred yards, 120°–300°; at depths of 83 ft (25 m) (northern peak), 52 ft (16 m) (central peak), and 69 ft (21 m) (southern peak).

Marisla Seamount was named after dive-cruise ship *Marisla II* (Mexican Flag), formerly USCG Cutter *Columbine*, owned by Maria Luisa Adcock and Richard M. Adcock. Richard was the first known sport diver, using SCUBA gear, to dive on the Seamount in 1957. Adcock began making commercial sport diving cruises to the sea mount utilizing *Marisla* (a converted LCM 56) and continued the dive business with *Marisla II* from 1968 through 2009. *Marisla* and *Marisla II* have both been scrapped.

4.28.1 External links

- In La Paz

- Scuba Travel

- The Seamount

- Baja Diving

4.29 Moai (seamount)

This article is about the submarine volcano. For other uses, see Moai (disambiguation).

The **Moai Seamount** is a submarine volcano, the second most westerly in the Easter Seamount Chain or Sala y Gómez ridge. It is east of Pukao seamount and west of Easter Island. It rises over 2,500 metres from the ocean floor to within a few hundred metres of the sea surface.[*][2] The Moai seamount is fairly young, having developed in the last few hundred thousand years as the Nazca Plate floats over the Easter hotspot.

The Moai seamount was named after the moai statues of neighbouring Easter Island.

4.29.1 See also

- Easter Island

- Sala y Gómez

4.29.2 References

[1] Geographic.org

[2] Haase, Karsten M.; Peter Stoffers; C. Dieter Garbe-Schönberg (October 1997). "The Petrogenetic Evolution of Lavas from Easter Island and Neighbouring Seamounts, Near-ridge Hotspot Volcanoes in the SE Pacific". *Journal of Petrology* **38** (06): 785–813. doi:10.1093/petrology/38.6.785. Retrieved 2010-03-16.

4.30 Monowai Seamount

Monowai is a volcanic seamount to the north of New Zealand. It is one of the most active volcanoes in the Kermadec volcanic arc and has erupted many times since 1977.[*][2][*][3] Coordinates: 25°53′13″S 177°11′17″W / 25.887°S 177.188°W

4.30.1   Description

The summit of Monowai is approximately 132 metres (433 ft)[*][1] below sea level, considerably above the level of the nearby Tonga and Kermadec Trenches. The summit's position and depth changed between 1998 and 2004, due to a landslide and eruptive regrowth. A 1,500 metres (4,900 ft) deep caldera, 13 by 8 km (8.1 by 5.0 mi), lies 5 to 15 km (3.1–9.3 mi) NNE of the seamount's main cone.

Monowai was discovered by an aerial survey in 1944. Subsequent surveys showed evidence of significant change - sea discolouration and seismic activity. Studies between 1978 and 2007 showed the summit repeatedly rising and falling.[*][4]

4.30.2   2010s

Sonar data from the research ship *R/V Sonne* showed that between May 14 and June 1–2, 2011 the summit collapsed by as much as 18.8 m (62 ft) and lava flows had raised another area by 79.1 m (260 ft). Additionally a new volcanic cone was created. The *R/V Sonne* also observed the sea to be a yellowy-green with gas bubbles. Researchers believe that only Vesuvius and Mount St Helens have recorded larger growth rates.[*][4]

In August 2012, a 25,000 km^2 (9,700 sq mi) floating raft of pumice was found offshore from New Zealand. It was initially believed to be from Monowai,[*][5][*][6] but Monowai was later ruled out as a possible source.[*][7]

4.30.3   See also

- List of volcanoes in New Zealand

4.30.4   References

[1] Global Volcanism Program: Monowai Seamount

[2] Global Volcanism Program: Monowai Seamount, eruptive history

[3] David Shukman (13 May 2012). "Rise and fall of underwater volcano revealed". *BBC News*. Retrieved 2012-05-13.

[4] BBC, News, Science - Environment, Rise and fall of underwater volcano revealed. by David Shukman

[5] Field, Michael; Kirk, Stacey (8 August 2012). "25,000 sq km sea of pumice floats off New Zealand". Stuff.co.nz. Retrieved 2012-08-10.

[6] "Vast volcanic 'raft' found in Pacific, near New Zealand". BBC News. 10 August 2012. Retrieved 2012-08-10.

[7] Cooke, Michelle (11 August 2012). "Scientists rock theory on pumice raft". *Stuff.co.nz*. Retrieved 13 August 2012.

4.30.5   External links

- NOAA Ocean Explorer: Monowai

- Global Volcanism Program: Monowai Seamount

- Volcanic Activity Reports for Monowai Seamount, 2007

- Collapse and Re-growth of Monowai Submarine Volcano, Kermadec Arc, 1998-2004, Chadwick et al., 2005.

4.31 Myōjin-shō

Myōjin-shō (明神礁) is a submarine volcano located about 450 kilometers south of Tokyo on the Izu-Ogasawara Ridge in the Izu Islands. Volcanic activity has been detected there since 1869. Since then it has undergone more eruptions, the most powerful of which resulted in the appearance and disappearance of a small island.

The name Myōjin-shō derives from a fishing boat, *No.11 Myōjin-Maru* of Yaizu City, Shizuoka Prefecture, the crew of which first witnessed the major volcanic eruption of 1952.

4.31.1 Eruption of 1952-3

The volcanic eruption from 1952 to 1953 was one of its biggest eruptions on record, with the repetitious appearance and disappearance of an island, which at one point reached over ten metres above sea level, before sinking after a major volcanic eruption in September 1953. On September 24, 1952, a survey vessel, *Kaiyo Maru No. 5* of the Hydrographic Department of the Maritime Safety Agency, was destroyed by the volcano, with the loss of its crew of 31 (including the nine scientists studying the eruption). Consequently, the Department developed *Manbou* (Sunfish), an unmanned radio operating survey boat, and has used it for the research of dangerous sea areas such as submarine volcanoes.

This was the first time that volcanic activity had been detected using the SOFAR channel.[*][1]

4.31.2 Survey of 1998-9

In 1998 and 1999, the Hydrography Department conducted comprehensive sea bottom surveys around Myōjin-shō, using the state-of-the-art survey vessel *Shoyo* and *Manbou II*, the second generation *Manbou*. As a result of these surveys, a detailed picture of the seabed topography around Myōjin-shō was made for the first time.

Manbou II conducted the survey of the sea area within a radius of 3 nautical miles (about 5.4 kilometers) of Myōjin-shō. *Shoyo* conducted the survey of the sea area within a radius of about 10 nautical miles (about 18.5 kilometers) but farther than the area of the radius of 3 nautical miles (5.6 km). *Manbou II* works by the order of preprogrammed instructions and measures depth and water temperature. Bathymetric survey of *Manbou II* was carried out by using the "PRD-601" echo sounder at intervals of 0.2 nautical miles (about 370 meters). *Shoyo* conducted a comprehensive survey including the geological and geophysical surveys of sea bottom. Bathymetric survey of *Shoyo* was carried out by using a "Seabeam 2112" echo sounder at intervals of 0.5 nautical miles (about 930 meters).

4.31.3 Structure

Previously, Myōjin-shō was considered to be the central cone of a double volcano with the Bayonnaise Rocks (rocks of 9.9 meters in height above the sea level) as a portion of the somma (Mita, 1949). As a result of the survey, however, the authors found that both Myōjin-shō and the Bayonnaise Rocks are cones on the somma of a double volcano. The foot of this double volcano lies 1,400 to 1,500 meters in depth and the size is about 30 by 25 kilometers east-west, north-south. The somma is almost a circle in the diameter of 7 by 9 kilometers and the height is 1,000 – 1,400 meters.

The diameter of the caldera floor is 5.6 kilometers and about 1,100 meters in depth. The central cone is a high formerly known as Takane-shō, 328 metres below sea level.

Myōjin-shō is a post caldera cone formed in the northeastern part of the somma of the double volcano. It is a single conical cone and its height is 550 meters with the shallowest depth 50 meters. A record that suggests a gushing of bubbles near the summit was obtained and micro-earthquakes were observed near Myōjin-shō, showing that the volcano is still active, although at a low level.

4.31.4 References

[1] http://www.gsajournals.org/perlserv/?request=get-abstract&doi=10.1130%2F0016-7606(1954)65%5B941%3ATDOMVE%5D2.0.CO%3B2 "Transpacific detection of Myojin volcanic explosions by underwater sound"

4.31.5 External links

- Survey of Myojin-sho

- "Bayonnaise Rocks". *Global Volcanism Program*. Smithsonian Institution.

4.32 Oshawa Seamount

The **Oshawa Seamount** is a seamount located in the Pacific Ocean off the coast of the Queen Charlotte Islands, British Columbia, Canada.

4.32.1 References

- British Columbia Marine Topography

4.32.2 See also

- Volcanism of Canada

- Volcanism of Western Canada

- List of volcanoes in Canada

4.33 Pactolus Bank

Pactolus Bank or **Burnham Bank** was discovered at 56°36′S 74°20′W / 56.600°S 74.333°W and a depth of 67 fathoms (123 m; 402 ft) by Captain W.D. Burnham on the American ship *Pactolus* on November 6, 1885.

Felix Riesenberg, who served under Burnham, postulated that Pactolus Bank was the sunken location of Elizabeth Island, discovered by Sir Francis Drake's ship the *Golden Hinde* in 1578. Leaving the Magellan Strait, Drake's ship was driven far to the west and south, before clawing its way back towards land. On 22 October the ship anchored off an island which Drake, according to Riesenberg, named "Elizabeth Island," where wood and water was collected and seals and penguins captured for food, along with "herbs of great virtue." According to Drake's Portuguese pilot, Nuño Da Silva, their position at the anchorage was 57°S. However, no island has been confirmed at that latitude, although a sinking volcanic island may have been sighted in that vicinity by the Danish ship *Lutterfeld* in December 1876 or 1877.

The USS Bear of Oakland investigated the area in March 1940 but her log does not mention any soundings or results. The USS *Wyandot* investigated the area in 1956 and found no indication of a shoal. The importance of these attempts is extremely limited, mainly due to the severe weather these ships had to operate in. So Pactolus Bank may itself be a phantom island.

In 2008 Dutchman Hylke Tromp, after some years of research, published a comprehensive report, containing all known facts and myths regarding the Pactolus Bank.

4.33.1 Sources

- Knox-Johnston, Robin (1994). *Cape Horn: A Maritime History*. London: Hodder & Stoughton. ISBN 0-340-41527-4., pp 40–45

- Riesenberg, Felix (1939). *Cape Horn: The Story of the Cape Horn Region*. New York: Dodd, Mead & Co. ISBN 1-881987-04-3.

- Stommel, Henry (1984). *Lost Islands: The Story of Islands That Have Vanished from Nautical Charts*. Vancouver: University of British Columbia Press. ISBN 0-7748-0210-3., pp 77–78

- Tromp, Hylke. *The Story of the Pactolus Bank. A lost Island near Cape Horn*. Private publication, 2008.

4.33.2 External links

- Pactolus Bank or Elizabeth Island? (Spanish)

- Tromp, Hylke. The Story of the Pactolus Bank. A lost Island near Cape Horn

4.34 Panov Seamount

Panov is minor seamount in the southeast Pacific located near the western part of the Valdivia Fracture Zone.

4.35 Pasco banks

The **Pasco Banks** refers to a naturally occurring geological and marine formation in the south Pacific Ocean.[1] The Pasco Banks is a long ridge-like seamount that rises from about 200 m to within 30 m of the ocean's surface. Covered in patchy coral reef, it attracts large schools of baitfish, mainly rainbow runner, which in turn are preyed upon by larger predatory fishes. This abundance of fish has made the Pasco Banks a popular and reliable fishing location for hundreds of years.

4.35.1 Location

The Pasco Banks seamount formation is located 13° 05' 00" S latitude and 174° 25' 00" W longitude in the South Pacific Ocean, roughly halfway between the Polynesian islands of Samoa and Uvea.[2] The seamount that includes the Pasco Banks is part of the Samoan volcanic chain which extends from Tuvalu, Uvea, and Futuna eastward to the submarine volcano Vailulu'u east of the Manu'a islands of American Samoa. The Samoan volcanic chain results from the plume-driven tectonic Samoa hotspot, not unlike the hotspot responsible for the formation of the Hawaiian Islands.[3]

4.35.2 Discovery and Prehistoric Fishing

Seafaring Austronesians, perhaps Lapita or proto-Polynesian migrants, were the first humans to discover and utilize the Pasco Banks. In pre-colonial times, Samoan master fishermen ("tautai" in the Samoan language) routinely led fleets of double-hulled voyaging canoes to the Pasco Banks for fishing expeditions.[4] Smaller canoes used for trolling were transported on the decks of the larger double-hulled canoes.[5] These expeditions of more than 80 miles of open sea to an entirely submerged reef attest to the maritime expertise of Oceanian navigators in developing the practical skill and scientific knowledge of Polynesian navigation. Samoan fishermen trolled for the same predatory fish as modern sport fishermen, including "atu" (skipjack tuna), "asiasi" (yellowfin tuna), "paala" (wahoo), "tagi" (dogtooth tuna), and "sa'ula" (marlin). Samoan fishermen also used scoop nets to harvest from the multitudes of baitfish at the Pasco Banks, especially "samani" (rainbow runner).

4.35.3 Historical Fishing

While Polynesians once fished the Pasco Banks regularly, the area is largely devoid of commercial fishing ventures today. The Samoan Fisheries Department reports several fishing ventures at the Pasco Banks beginning in 1979 but the remote location and concerns about profitability and sustainability resulted in most projects being cancelled by 1985.[6]

4.35.4 Modern Sport Fishing

Fishing expeditions to the Pasco Banks are often arranged through private chartered outfits based in Samoa and Vanuatu.*[7] The most abundant catches are reported to be Dogtooth tuna, wahoo, skipjack tuna, marlin, yellowfin tuna and sailfish.

4.35.5 References

[1] 1980. Natland, James. "The Progression of Volcanism in the Samoan Linear Volcanic Chain," American Journal of Science, 280(A):709-735

[2] National Geospatial Intelligence Agency; http://www.geographic.org/geographic_names/name.php?uni=$-$240774&fid=6440& c=undersea_features

[3] 2004. Hart et al. "Genesis of the Western Samoa Seamount Province," Earth & Planetary Science, 227:37-56.

[4] 1895. Stair, J.B. "Flotsam & Jetsam. . . Early Samoan Voyages," Journal of the Polynesian Society, 4:99-131.

[5] 1995. Kramer, Augustin. "The Samoa Islands, Vol. II," Polynesian Press

[6] 1987. Robert Gillet & Taniela Sua. "FAO/UNDP Regional Fishery Support Programme, Field Document 87/6

[7] 2009. Cooper, Steve. "Vanuatu offers affordable game fishing with reliable returns for your effort," Sunday Herald Sun

4.36 Peirce Seamount

Peirce Seamount, also called **Pierce Seamount**, is a seamount located in the Pacific Ocean west of the Queen Charlotte Islands, British Columbia, Canada. It lies between Denson Seamount and Hodgkins Seamount and is member of the Kodiak-Bowie Seamount chain, a chain of seamounts in southeastern Gulf of Alaska stretching from the Aleutian Trench in the north to Bowie Seamount in the south.

4.36.1 See also

- List of volcanoes in Canada

- Volcanism of Canada

- Volcanism of Western Canada

4.36.2 References

- British Columbia Marine Topography

4.36.3 External links

- Seamount Catalog -- Peirce Seamount

4.37 Pioneer Seamount

Pioneer Seamount is an undersea mountain, or seamount, in the Pacific Ocean off the coast of central California.

4.37.1 Location

Pioneer Seamount is located at 37° 21.1' North Latitude, 123° 26.1' West Longitude,*[1] at the base of the continental slope*[2] of North America about 95 kilometers (59 miles) off the coast*[1] just southwest of San Francisco, California.*[2]

4.37.2 Physical characteristics

The seamount is a volcano between 10.9 and 11.1 million years old. It is about 12.8 kilometers (8.0 miles) long as well as about 12.8 kilometers (8.0 miles) wide, and has a volume of about 135 cubic kilometres (32 cubic miles). It rises about 1,930 meters (6,330 feet) above the surrounding ocean floor, and its peak is a minimum of 820 meters (2,690 feet) below the ocean 's surface. Samples from the seamount consist of highly vesicular alkalic basalt, hawaiite, and mugearite.*[3]

4.37.3 Biological environment

A wide variety of sealife lives on the seamount. Corals dominate in deeper areas and sponges in its shallower parts.*[4]

4.37.4 Naming

Pioneer Seamount was named for *Pioneer*, the first of three survey ships of the United States Coast and Geodetic Survey to bear the name. *Pioneer* operated along the United States West Coast and in the then-Territory of Alaska during her Coast and Geodetic Survey career, which lasted from 1922 to 1941.*[5]

4.37.5 Notes

[1] NOAA Vents Program Acoustic Monitoring; Acousting Monitoring Program Pioneer Seamount

[2] United States Geological Survey Monthly Newsletter *Sound Waves*: Fieldwork: Geologists and Biologists Endeavor to Understand Seamount Environments Off California

[3] Monterey Bay Aquarium Research Institute: Pioneer Seamount

[4] Monterey Bay Aquarium Research Institute: Seamounts 2007

[5] NOAA History: Tools of the Trade: Coast and Geodetic Survey Ships: Pioneer

4.37.6 References

- NOAA Vents Program Acoustic Monitoring; Acousting Monitoring Program Pioneer Seamount

- United States Geological Survey Monthly Newsletter *Sound Waves*: Fieldwork: Geologists and Biologists Endeavor to Understand Seamount Environments Off California

- Monterey Bay Aquarium Research Institute: Pioneer Seamount

- NOAA History: Tools of the Trade: Coast and Geodetic Survey Ships: Pioneer

4.38 President Jackson Seamounts

The **President Jackson Seamounts** are a series of seamounts (underwater volcanoes) located on the Pacific Plate, off of California.*[2] It consists of 8 seamounts, 4 independent and 4 morpohologically fused, just west of the northern Gorda Ridge. They are generally very small, and arranged linearly.

The assembly consists of 8 seamounts arranged in a chain. The easternmost of the seamounts is located about 53 km (33 mi) east of the axial ridge that spawned the volcanoes, Gorda Ridge. The chain was mapped initially with SeaBeam equipment in 1985, however its incomplete coverage limited its usefulness. A GLORIA sidescan survey conducted in 1986 further established the rough size and locations of the seamounts. A further sampling cruise by the United States Geological Survey (USGS) recovered pillow lava and hyaloclastite from 4 of the 8 volcanoes. The samples seemed to be of more primitive origin then those erupted elsewhere on the ridge. In 1998, the Monterey Bay Aquarium Research Institute conducted two dives on a cone cluster located southeast of the main structure.[1]

The chain consists of eight volcanoes. Of these, four are self-dependent, whereas the other four form twin fused, morphalogically complex, flat-topped structures. The volcanoes form a linear chain, and are relatively small, measuring less than 10 km (6 mi) in diameter, and are nearly circular in shape. They average 47 km^3 (11 cu mi)±14 km^3 (3 cu mi) in volume, but range from 24 km^3 (6 cu mi) to 68 km^3 (16 cu mi). In addition to the eight main seamounts, there is a scattered disk of minor volcanic cones and flows about 13 km (8 mi) southeast of the chain.[1]

The summits of the seamounts, with the exception of the second one from the northwest, are all flat. Several have nested or cross-layered calderas and pit craters (29 in total), many of which push into the flank of the volcanoes. In several cases, newer volcanic structures have all but erased older ones. In addition to all this the seamounts are pocketed by multiple small cones, and are littered by debris ranging far and wide from the range. There is evidence of erosional activity throughout, with erosional valleys and landslide debris visible.[1]

4.38.1   See also

- Vance Seamounts

4.38.2   References

[1] "Mapping Program: President Jackson Seamounts". Monterey Bay Aquarium Research Institute. Feb 6, 2009. Retrieved 28 December 2009.

[2] "Seamount Catalog". *Seamount database*. Earthref.org, a National Science Foundation project. Retrieved 28 December 2009.

4.39   Pukao (seamount)

The **Pukao Seamount** is a submarine volcano, the most westerly in the Easter Seamount Chain or Sala y Gómez ridge. To the east are Moai (seamount) and then Easter Island. It rises over 2,500 metres from the ocean floor to within a few hundred metres of the sea surface.[1] The Pukao Seamount is fairly young, and believed to have developed in the last few hundred thousand years as the Nazca Plate floats over the Easter hotspot.

4.39.1   See also

- Easter Island

- Sala y Gómez

4.39.2   References

[1] Haase, Karsten M.; Peter Stoffers and C. Dieter Garbe-Schönberg (October 1997). "The Petrogenetic Evolution of Lavas from Easter Island and Neighbouring Seamounts, Near-ridge Hotspot Volcanoes in the SE Pacific". *Journal of Petrology* **38** (6): 785–813. doi:10.1093/petrology/38.6.785. Retrieved 2010-03-16.

4.40 Rivadeneyra Shoal

Rivadeneyra Shoal is a shoal or seamount in the Eastern Pacific Ocean between Malpelo Island and Cocos Island. It was reported in October 1842 at the position 4°15′N 85°10′W / 4.250°N 85.167°WCoordinates: 4°15′N 85°10′W / 4.250°N 85.167°W with a depth of 10 feet. It was unsuccessfully searched for by the British war vessels *Cockatrice* and *Havannah* in 1854 and 1857 respectively, and by the USS *Mohican* in 1885.

4.40.1 See also

- 1891 German map showing Rivadinera Shoal

4.40.2 Sources

- Findlay, Alexander George (1851). *A Directory for the Navigation of the Pacific Ocean*. London: Laurie, p 1054.

- Great Britain Hydrographic Office (1860). *South America Pilot: Including Magellan Strait, the Falkland and Galapagos Islands*. Eyre and Spottiswoode, p 394.

- Stommel, Henry (1984). *Lost Islands: The Story of Islands That Have Vanished from Nautical Charts*. Vancouver: University of British Columbia Press, p 113. ISBN 0-7748-0210-3.

4.41 Rosa Seamount

Rosa Seamount is an uplifted piece of the sea floor west of the Baja California. It is possible it was created by a shift of land during an earthquake located at the Cedros Trench.

4.42 Schmieder Bank

Schmieder Bank is a rocky bank west of Point Sur, California, roughly 25 nautical miles (46 km) south of Monterey, supporting an extraordinarily lush biological community, including very large individual colonies of the California Hydrocoral [Schmieder, 1989].

The bank lies about 3 nm WSW offshore from Point Sur. Within the 75 m contour, the bank is roughly elliptical, with major axis NW-SE, enclosing an area of about 1 square mile (3 km^2). The surface of the bank is a surf-erosional plateau, punctuated by a series of narrow ridges (running approximately parallel to the major axis), and several extremely sharp isolated pinnacles. Minimum depths are 36 m (at one location) and 40 m (at least 4 locations). During significant ice ages the Bank emerged as an offshore island. The Bank is an example of an Underwater Island, of which Cordell Bank is another [Schmieder, 1991].

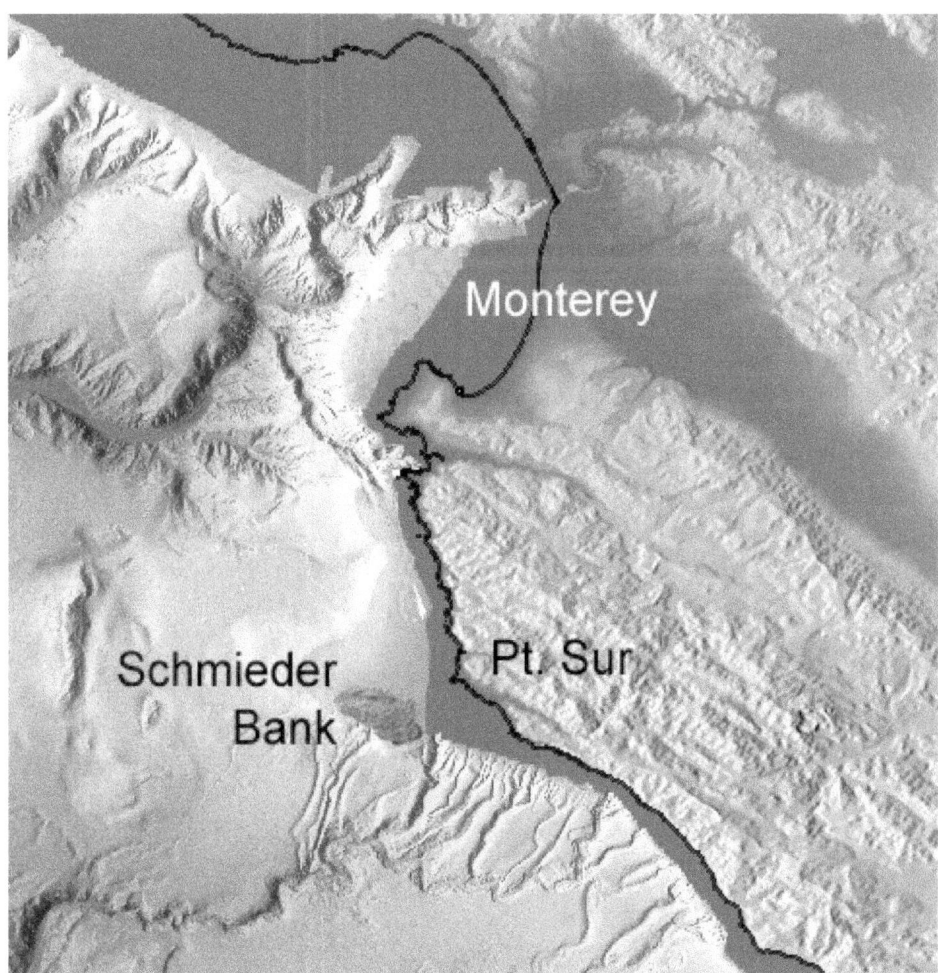

The Bank was first explored by divers during 1988-91 by Cordell Expeditions, organized and led by Dr. Robert Schmieder. That work generated a general description of the bank, summarized in a privately published report [Schmieder, 1989]. Because of the exceptionally rich biological community, the boundary of the Monterey Bay National Marine Sanctuary was extended to include the Point Sur area.[*][1] The Bank is similar in many respects to Cordell Bank [Schmieder, 1991], lying 121 nm to the Northwest. The Bank has become a desirable, but difficult, goal for SCUBA divers.[*][2][*][3][*][4]

During 1986, NOAA carried out a high-resolution multibeam survey of the area as part of the Exclusive Economic Zone (EEZ) program [NOAA, 1986], and during 1998 the Monterey Bay Aquarium Research Institute (MBARI) contracted to carry out a high-resolution (5 m horizontal) survey [USGS, 1989].

Schmieder Bank was named[*][5] on October 15, 1990, by the U. S. Board on Geographic Names, following a recommendations by the late Dr. Melanie Stright (U.S. Minerals Management Service), Dr. Paul Silva (UC Berkeley) and Dr. Sylvia Earle (NOAA Chief Scientist).

4.42.1   References

[1] http://montereybay.noaa.gov/.

[2] .

[3]

[4] .

[5]

- NOAA, 1983, Navigational Chart 18686.

- NOAA, 1986, EEZ Bathymetric Survey, Project No. OPR-M539-DA-86, Description Report, Field No. DA-20-3-86, Register B-94.

- Robert W. Schmieder, 1989, "The 1988 Expeditions to Pt. Sur. Summary of Results," Rept. CE-89-4, Cordell Expeditions, 80 pp. incl. color plates.

- Robert W. Schmieder, 1991, "Ecology of an Underwater Island", Cordell Expeditions, Walnut Creek, CA. http://www.cordell.org.

- USGS, 1989, Proceedings of the 1989 Exclusive Economic Zone Symposium on Mapping and Research: Federal-State Partners in EEZ Mapping, Millington Lockwood, Bonnie A. McGregor, Editors, Meetings at the USGS National Center, Reston, VA, Nov. 14-16, 1989. USGS Circular 1052.

4.43 Seminole Seamount

The **Seminole Seamount** is a seamount located in the Pacific Ocean off the coast of northern Vancouver Island, British Columbia, Canada.

4.43.1 References

- British Columbia Marine Topography

4.43.2 See also

- Volcanism of Canada

- Volcanism of Western Canada

- List of volcanoes in Canada

4.44 Siletz River Volcanics

The **Siletz River Volcanics**, located in the Oregon Coast Range, United States, is a sequence of basaltic pillow lavas that make up part of Siletzia. The basaltic pillow lavas originally came from submarine volcanoes that existed during the Eocene.

The Paleocene to Eocene volcanics consist of volcanism flows and sills of tholeitic to alkalic basalts with associated tuff-breccia, siltstone and sandstone. The flows are vesiculated with zeolite filled amygdules.

The volcanics originated as oceanic crust and seamounts. Potassium argon dating gives an age of 50.7 ± 3.1 to 58.1 ± 1.5 Ma.[*][1]

The sequence has been divided into a lower pillowed tholeiitic unit and an upper porphyritic alkali basalt unit.[*][2]

The volcanics occur in the following counties of western Oregon: Benton, Coos, Douglas, Lane, Lincoln, Polk, Tillamook, Washington and Yamhill.[*][1]

4.44.1 References

[1] "Siletz River Volcanics and related rocks". *USGS Mineral Resources On-Line Spatial Data*. U.S. Geological Survey.

[2] Snavely, Parke D.; MacLeod, Norman S.; Wagner, Holly C. (June 1968). "Tholeiitic and alkalic basalts of the Eocene Siletz River Volcanics, Oregon Coast Range". *American Journal of Science* **266** (6): 454–481. doi:10.2475/ajs.266.6.454.

4.45   South Chamorro Seamount

South Chamorro Seamount is a large serpentinite mud volcano and seamount located in the Izu-Bonin-Mariana Arc, one of 16 such volcanoes in the arc. These seamounts are at their largest 50 km (31 mi) in diameter and 2.4 km (1.5 mi) in height. Studies of the seamount include dives by the submersible dives (DSV *Shinkai*, 1993 and 1997), drilling (Ocean Drilling Program, 2001), and ROV dives (2003, 2009).[*][1]

The seamount and its nearby peers were created by the movement of crushed rock, resulting from plate movement, upwards through fissures in the Mariana Plate. South Chamorro is the farthest of the mud volcanoes from the trench, at a distance of 85 km (53 mi), resulting in high-temperature flows rich in sulfate and methane. The seamount suffered a major flank collapse on its southeastern side, over which the present summit was probably formed. The summit supports an ecosystem of mussels, gastropods, tube worms, and others, suggesting that it is an active seeping region.

4.45.1   Geology

Origin and geochemistry

South Chamorro Seamount and the other mud volcanoes formed as a result of the subduction of the Pacific Plate below the Mariana Plate; fault lines in the Mariana Plate provide a gateway for churned up rock and fluid from the grinding process up to the ocean surface. Reactions with the overlying mantle produce serpentinite, hydrogen gas, and other alkaline substances.[*][1]

South Chamorro Seamount was first recognized as a mud volcano is 1977, on the basis of sonar data, and confirmed as such in 1981 by the collection of serpentine and schist. is one of the farthest volcanoes from the trench, 85 km (53 mi) away, where the plate rides approximately 25 km (16 mi) underneath. Because of its distance from the trench, its eruptive fluids are hotter (over 350 °C or 662 °F), less alkalic, and with more calcium. Its flows have a pH of 12.2 (highly basic) and are sulfate and methane rich.[*][1]

Serpentine mud volcanoes are currently limited to the Izu-Bonin-Mariana arc; however, there is evidence of the geological remnants of similar volcanics globally, throughout Earth's history.[*][1]

Structure

Side-scan surveys of South Chamorro Seamount show a major southeastern sector edifice collapse, with debris flows of serpentine material (dredged in 1981 and observed by submersibles in 1995) that blankets the slope of the trench from summit to axis. The true summit of the volcano sits above this collapse; its formation was likely tied to the collapse. Submarine observation of the summit show that the summit knoll is broken up into slabs of serpentine mud, with meter-deep fissures arranged in a crosscutting orientation.[*][2]

4.45.2   Ecology

South Chamorro Seamount is host to a variety of fauna, including mussels, gastropods, tube worms, and galatheid crabs. A borehole observatory on the summit produced 20,000 litters of microbally-altered fluid per day for study.[*][1] These organisms are supported by low-temperature springs from the fissures surrounding the summit zone. The mussels are probably of the genus Bathymodiolus, which require a high concentration of methane in their food source. The fluid composition and the biological community suggests that the summit region is an active cold seep region.[*][2]

4.45.3   See also

- Graveyard Seamounts

- Jasper Seamount

- Mud volcano

- Muirfield Seamount

- Sedlo Seamount

4.45.4 References

[1] C. Geoffrey Wheat, Patricia Fryer, Ken Takai, Samuel Hulme. "Spotlight 9: South Chamerro Seamount" (PDF). *Oceanography*. Seamounts Special Issue (Oceanography Society) **23** (1). Retrieved 28 July 2010.

[2] "Site 1200: Serpentine Mud Volcano Geochemical Observatory: Geological Setting". Ocean Drilling Program. Retrieved 28 July 2010.

4.46 Stirni Seamount

The **Union Seamount** is a seamount located in the Pacific Ocean off the coast of northern Vancouver Island, British Columbia, Canada.

4.46.1 References

- British Columbia Marine Topography

4.46.2 See also

- Volcanism of Canada

- Volcanism of Western Canada

- List of volcanoes in Canada

4.47 Suiyo Seamount

Suiyo Seamount is a seamount (submarine volcano) off the eastern coast of Japan, directly south of Torishima and Sofugan volcano at the southern tip of the Izu Islands. The volcano is one of the Shichiyo Seamounts, a small group of submarine volcanoes named after different days of the week ("Suiyo" means "Wednesday" in Japanese).[1]

Suiyo consists of a basaltic to dacitic submarine caldera and lava dome, and rises about 1,400 m (4,590 ft) from its base on the sea floor to within 1,418 m (4,652 ft) of the surface. Suiyo has a prominent summit caldera, 1.5 km (0.9 mi) wide and 500 m (1,640 ft) deep.[1]

The volcano's excised (weathered) structure suggests that it is of older age then some of the other volcanoes in the group. Suiyo is covered by a thick sediment cap, a feature that collects over a long span of inactivity, and fault patterns and valleys have been observed on its flanks.[2]

Suiyo Seamount is associated with a magnetic anomaly: ocean-floor surveys of it and the surrounding area found that a large negative rock body existed to the east of the seamount, while positive bodies existed to the northwest and south. The reasons for this complex anomaly, which also exists in several other nearby seamounts, is unknown, but is suggested to be the result of interactions between different magnetic fields of different ages.[2]

A burst of hydrothermal activity was observed in July 1991, raising water temperatures at the vent to 290 °C (550 °F); following the event, the volcano, until then thought extinct, was reclassified as active by the Japan Meteorological Agency.[1] A bathymetric survey of the volcano found sulfur-oxidizing microbes to be predominant, and concluded that Suiyo Seamount was a natural "incubator" for this bacterial type.[3]

4.47.1 References

[1] "Suiyo Seamount". *Global Volcanism Program*. Smithsonian National Museum of Natural History. Retrieved 10 October 2010.

[2] Makoto Yuasa, Fumitoshi Murakami, Eiji Saito, and Kazuaki Watanabe (1991). "Submarine topography of seamounts on the volcanic front of Izu-Ogasawara (Bonin) Arc" (PDF). *Bulletin of the Geological Society of Japan* (Geological Society of Japan) **42** (12): 703–743. Retrieved 10 October 2010.

[3] Michinari Sunamura, Yowsuke Higashi, Chiwaka Miyako, Jun-ichiro Ishibashi, and Akihiko Maruyama (February 2004). "Two Bacteria Phylotypes Are Predominant in the Suiyo Seamount Hydrothermal Plume". *Applied and Environmental Microbiology* (American Society for Microbiology) **70** (2): 1190–1198. doi:10.1128/AEM.70.2.1190-1198.2004. Retrieved 10 October 2010.

4.48   Supply Reef

Supply Reef is a submerged circular reef of volcanic origin in the Northern Mariana Islands chain, about 10 kilometres (6 mi) NW of the Maug Islands. Presently this igneous seamount is roughly 8 metres (26 ft) below the ocean's surface and about 100 m (300 ft) in diameter.

4.48.1   References

• "Supply Reef". *Global Volcanism Program*. Smithsonian Institution.

4.49   Taney Seamounts

The **Taney Seamounts** are a range of five extinct underwater volcanoes located 300 kilometres (160 nmi; 190 mi) west of San Francisco on the Pacific Plate.*[1] The seamounts were identified during the United States Geological Survey's scan of the Exclusive Economic Zone,*[2] conducted in the 1980s with the GLORIA sidescan sonar.*[3]

4.49.1   References

[1] "Mapping Program: Taney Seamounts". Monterey Bay Aquarium Research Institute. Feb 6, 2009. Retrieved 29 December 2009.

[2] Petersen, J.K. (2007). *Understanding Surveillance Technologies: Spy Devices, Privacy, History, & Applications, Revised and Expanded Second Edition*. CRC Press. p. 228. ISBN 978-0-8493-8320-5.

[3] "U.S. Exclusive Economic Zone (EEZ) GLORIA Mapping Program". US Geological Survey. Retrieved 6 November 2010.

4.50   Tasmanian Seamounts

The **Tasmanian Seamounts** (also *Tasman Seamounts* and *Tasmania Seamounts**[4]) are a group of seamounts (underwater volcanoes) located off the southern tip of Tasmania. The seamounts were created more than 55 million years ago by the Tasman hotspot. The seamounts are ecologically important, and harbor a lush marine ecosystem, but are threatened by overfishing. For this reason, part of the Tasmanian Seamounts were incorporated into a marine reserve in 1999.

4.50.1   Geology

The Tasmanian Seamounts were created by the Tasman hotspot, a 4,000 km (2,000 mi) long mantle plume that is currently the active center of Mount Erebus in Antarctica.*[3] The seamounts, created roughly 55 million years ago, are between 1,000 and 2,000 m (3,000 and 7,000 ft) deep, 25 km (16 mi) across, and 200 to 500 m (700 to 1,600 ft) tall.*[2]

4.50.2   Ecology

The Tasmanian Seamounts are an important feature of the south Tasmanian marine environment. While oceans generally contain few nutrients, the presence of seamounts increases the flowing speed of the water current. This effect, created by the topography of the seafloor mountain, clears rocks in the area of sediments and provides food for filter feeders. The swept rocks serve as substrate for sponges and corals to attach to. Among them are some of the longest–living organisms on Earth, reaching an age of "hundreds and possibly thousands of years".*[5] The Tasmanian Seamounts in particular are dominated by the stony coral *Solenosmilia variabeles*; these corals provide a roost and living space for benthic mollusks, echinoderms and crustaceans like stone crabs (*Parclamis sp.*), squat lobsters (*Munidopsis treis*), and top snails (*Trochidae sp.*), thus solidifying the seamounts' ecology. Although many of these species are also found on the continental shelf, they are never found as concentrated or lush as they are on the seamounts.*[2]

The seamounts have been the site of commercial trawling for orange roughy for decades. In 1994, the Australian Geological Survey Organization mapped the south Tasmanian seafloor, including the Tasmanian Seamounts; shortly thereafter, Environment Australia (now the Department of Sustainability, Environment, Water, Population and Communities) released a report stating that the Tasmanian Seamount fauna was highly diverse, inadequately studied, and vulnerable to overfishing, and recommended that an oceanic reserve be created on the site. An interim three–year ban on trawling was declared to allow a study of the seamounts to be carried out in 1997 which found 242 distinct species of invertebrates. 26% to 44% of them could not be identified and are thus believed to be new, and about a third are seemingly present only in the seamount environment. Damage from trawling was also examined, and on the most heavily affected seamounts it had destroyed coral aggregations, removed 46% of species, and reduced net biomass by over half.*[1]

Following the study, the proposed marine reserve, the *Tasmanian Seamount Community* or *Tasmanian Seamounts Marine Reserve* was created as part of the Environment Protection and Biodiversity Conservation Act 1999.*[4] The reserve banned fishing, especially trawling, within an area of 370 km^2 (143 sq mi). It contained 70 seamounts, a fifth of the known Tasmanian seamounts, all of which were low-trawl areas and thus still in relatively unaffected form.*[6] In a 2007 audit it was merged into the Huon Commonwealth Marine Reserve.*[5] A more detailed bathymetric survey the same year found a total of 123 seamounts within the reserve, many of which were previously unknown.*[2]

4.50.3   References

[1] J. A. Koslow, K. Gowlett-Holmes, J. K. Lowry, T. O' Hara, G. C. B. Poore, A. Williams (April 4, 2011). "Seamount benthic macrofauna off southern Tasmania: community structure and impacts of trawling" (PDF). *Marine Ecology Progress Series* **213**: 111–125. Retrieved 27 October 2011.

[2] "Seamount ecosystems conserved in the Huon Commonwealth Marine Reserve" (PDF). June 2007. Retrieved 26 October 2011.

[3] Seach, John. "Tasman Seamounts Reserve - John Seach". Volcanolive. Retrieved 27 October 2011.

[4] "Tasmanian Seamounts". Department of Sustainability, Environment, Water, Population and Communities. 20 June 2011. Retrieved 27 October 2011.

[5] "Huon Commonwealth Marine Reserve". Department of Sustainability, Environment, Water, Population and Communities. 29 June 2011. Retrieved 27 October 2011.

[6] "Fact Sheet: Discovering Seamounts". CSIRO Marine and Atmospheric Research. December 11, 2008. Retrieved 27 October 2011.

Coordinates: 44°18′S 147°00′E / 44.300°S 147.000°E

4.51 Teahitia

Teahitia is a submarine volcano, located 40 km (25 mi) northeast of the southeast tip of Tahiti of the Society Islands in the Pacific Ocean, with its peak 1600 meters below the water surface. It belongs to the Teahitia-Mehetia hotspot.

Teahitia's last eruption occurred in 1985.*[1]

4.51.1 References

[1] Global Volcanism Program: Teahitia

Coordinates: 17°34′S 148°51′W / 17.57°S 148.85°W

4.52 Three Wise Men (volcanoes)

The **Three Wise Men** are a row of three seamounts (underwater volcanoes) located in the Pacific Ocean, on the East Pacific Rise. They are part of a large group of seamounts, collectively known as the Rano Raji. They stand at between 1,000 m (3,281 ft) and 2,000 m (6,562 ft), and are named after the Biblical Magi or the "three wise men". The middle of the three is the tallest and also the flattest at its top. The southern one is similar to its larger neighbor, but slightly shorter. The northern one is the middle of the two, with a large caldera and a circular shape.

4.52.1 References

- Donna O'Meara (2008). "5: Volcanoes and the Sea". *Volcano: A Visual Guide* (1 ed.). P.O. Box 1338, Elicott Station Buffalo, New York 14205: Firefly Books. p. 246. ISBN 978-1-55407-353-5.

4.53 Tucker Seamount

The **Tucker Seamount** is a seamount located in the Pacific Ocean off the coast of northern Vancouver Island, British Columbia, Canada.

4.53.1 References

- British Columbia Marine Topography

4.53.2 See also

- Volcanism of Canada

- Volcanism of Western Canada

- List of volcanoes in Canada

4.54 Tuzo Wilson Seamounts

The **Tuzo Wilson Seamounts**, also called **J. Tuzo Wilson Knolls** and **Tuzo Wilson Knolls**, are two young active*[1] submarine volcanoes off the coast of British Columbia, Canada, located 200 km (124 mi) northwest of Vancouver Island and south of the Queen Charlotte Islands. The two seamounts are members of the Kodiak-Bowie Seamount chain, rising

500 m (1,640 ft) to 700 m (2,297 ft)*[2] above the mean level of the northeastern Pacific Ocean and is a seismically active site southwest of the southern end of the Queen Charlotte Fault. They are named after Canadian geologist John Tuzo Wilson.*[3]

4.54.1 Geology

The two submarine volcanoes are capped by hawaiite and are surrounded by numerous smaller vents, with a total edifice volume of about 12 km^3.*[2]

The lava emitted in eruptions at the Tuzo Wilson Seamounts is made of basalt, a common gray to black or dark brown extrusive volcanic rock low in silica content (the lava is mafic) that is usually fine-grained due to rapid cooling of lava. Glassy pillow lava is found at the seamounts, a type of rock typically formed when basaltic lava emerges from a submarine volcanic vent. The viscous lava gains a solid crust on contact with the water, and this crust cracks and oozes additional large blobs or "pillows" as more lava emerges from the advancing flow.

The origin of the Tuzo Wilson Seamounts is not without controversy. Some geologists theorize that the Tuzo Wilson Seamounts are linked with a hotspot because lava at the Tuzo Wilson Seamounts are fresh, glassy pillow basalts of recent age, as expected if these seamounts are located above or close to a hotspot south of the Queen Charlotte Islands.*[3]*[4] Others prefer rifting as the cause of volcanism because the seamounts are close to the Explorer spreading center.*[1] No theory is close to airtight. Part of the controversy is due to the uncertain origin of the Kodiak-Bowie Seamount chain. There is a 360 km (224 mi) long gap between recently (Late Pleistocene to Holocene) active Bowie and Tuzo Wilson Seamounts, both of which have erupted alkaline basalts of similar composition. If a mantle plume was responsible for activity at both seamounts, then it is likely that there would be evidence for alkaline volcanic activity in the area between these two seamounts.*[1]

4.54.2 See also

- Volcanism of Canada
- Volcanism of Western Canada
- List of volcanoes in Canada
- Geology of the Pacific Northwest

4.54.3 References

[1] A near-ridge origin for seamounts at the southern terminus of the Pratt-Welker Seamount Chain, northeast Pacific Ocean Retrieved on 2008-03-07

[2] "The Tuzo Wilson Volcanic Field, NE Pacific: Alkaline volcanism at a complex, diffuse, transform-trench-ridge triple junction". Retrieved 2007-08-13.

[3] J. Tuzo Wilson Knolls: Canadian hotspot Retrieved on 2007-08-13

[4] Origin of Igneous Rocks: The Isotopic Evidence Retrieved on 2008-03-08

4.54.4 External links

- Map of Tuzo Wilson Seamounts (seamount), Pacific Ocean

4.55 Union Seamount

The **Union Seamount** is a seamount located in the Pacific Ocean off the coast of northern Vancouver Island, British Columbia, Canada.

4.55.1 References

- British Columbia Marine Topography

4.55.2 See also

- Volcanism of Canada

- Volcanism of Western Canada

- List of volcanoes in Canada

4.56 Vailulu'u

Vailulu'u is a volcanic seamount discovered by geophysicist Rockne Johnson in the Samoa Islands on October 18, 1975.[*][1] The finding of an active, undersea, hotspot volcano is significant for scientists studying the Earth's fundamental processes.[*][2]

In size and appearance, Vaululu'u resembles Mount Fuji and ranges more than 33 km across the ocean floor at its base.[*][2] It rises 4200 m from the sea floor to a depth of 590 m. It is located roughly one-third of the way between Ta'u and Rose islands at the eastern end of the Samoa hotspot chain. The hotspot chain includes American Samoa and the Independent State of Samoa and extends west to the islands of Uvea or Wallis Island (Wallis and Futuna) and Niulakita (Tuvalu).[*][3]

The basaltic seamount is considered to mark the current location of the Samoa hotspot. The summit of Vailulu'u contains a 2 km wide, 400 m deep oval-shaped caldera. Two principal rift zones extend east and west from the summit, parallel to the trend of the Samoan hotspot. A third less prominent rift extends southeast of the summit.

The rift zones and escarpments produced by mass wasting phenomena give the sea mount a star-shaped pattern. On July 10, 1973, explosions from Vailulu'u were recorded by SOFAR (hydrophone records of underwater acoustic signals). An earthquake swarm in 1995 may have been related to an eruption from the seamount. Turbid water above the summit shows evidence of ongoing hydrothermal plume activity.

Evidence released in 2006 suggests that Vailulu'u may breach the surface of the ocean and officially become an island during this century.[*][4]

The name Vailulu'u comes from a Samoan story related to the sacred sprinkling of rain associated with gatherings for the Tui Manu'a paramount chiefs of Manu'a Island. The name came from an American Samoa student Taulealo Vaofusi, the winner of a naming competition.[*][2]

4.56.1 Nafanua

Nafanua is an active underwater volcanic cone that has been growing inside the summit crater of Vailulu'u since 2001. In 2005 it was 300 m tall, but still 708 m below sea level.[*][5] It is best known as the site of 'Eel City,' a hydrothermal vent biological community consisting mainly of synaphobranchid eels *Dysommina rugosa* (rather than the usual invertebrates).[*][6]

Around the base of Nafanua has formed a "Moat of Death" - a toxic zone formed by fluids emanating from Nafanua.[*][7] The ground here is covered with carcasses of dead fish, squids, crustaceans.

4.56.2 Gallery

- The summit of Nafanua is covered with thick microbial mats, indicative of low-temperature venting

- Broken pillow lavas, colored red by iron oxide, inside Vailulu'u crater.

- An octopus living on the western summit of Vailulu'u

- Swimming elasipod sea cucumber, *Paleopatides sp.*, photographed off the northern shore of Tau Island, Vailulu'u Expedition 2005

4.56.3 See also

- Nafanua, a war goddess in Samoan mythology.

4.56.4 References

[1] "Vailulu'u undersea volcano: The New Samoa" (PDF). *G3, An Electronic Journal of the Earth Sciences, American Geophysical Union*. Research Letter, Vol. 1. Paper number 2000GC000108. 8 December 2000. ISSN 1525-2027. Retrieved 2 December 2009. |first1= missing |last1= in Authors list (help)

[2] Lippsett, Laurence (3 September 2009). "Voyage to Vailulu'u". *Woods Hole Oceanographic Institution*. Retrieved 2 December 2009.

[3] Samoan Hotspot Trail at the Wayback Machine (archived December 23, 2010)

[4] Foxnews.com Emerging Ocean Volcano, Retrieved 30 August 2007

[5] Staudigel, H. et al. (2006) *Vailulu' u Seamount, Samoa: Life and death on an active submarine volcano*, Proceedings of the National Academy of Sciences, volume 103 Number 17, pages 6448 to 6453, Retrieved 30 August 2007

[6] Astrobiology Magazine: Extremes of Eel City Retrieved 30 August 2007

[7] "Vailulu'u Eel City and Moat of Death". Wondermondo.

- "Vailulu'u". *Global Volcanism Program*. Smithsonian Institution. Retrieved 2009-04-29.

4.56.5 External links

- NOAA Ocean Explorer Vailulu'u 2005 Expedition, Retrieved 30 August 2007

- Vailulu'u web site, information about the Vailulu'u seamount

4.57 Vance Seamounts

The **Vance Seamounts** are a group of seven submarine volcanoes located west of the Juan de Fuca Ridge. Most of the seamounts contain a caldera. They are the southernmost of several near-ridge chains located on the Pacific Plate, stemming from the Juan de Fuca Ridge.[*][1]

The easternmost five of the seven were surveyed using SeaBeam bathymetry. A chemical analysis of the lavas of the volcanoes indicated that the materials erupted from the volcanoes are relatively primitive. With the exception of one, the volcanoes form a nearly linear chain of circular volcanic cones. The second volcano from the northwest is not discretely shaped like the rest, rather it is a rough structure of lava cones and flows, which cover an area of at least 25 km × 10 km (16 mi × 6 mi). The six conical volcanoes (not including the odd one) range from 15 to 67 km^3 (4 to 16 cu mi) in volume, averaging 34 ± 15 km^3 (8 ± 4 cu mi).[*][1]

Most of the volcanoes have gently sloping plateaus, in conformity with their low shield volcano–like profiles, which have been heavily modified by multiple calderas. These are generally very shallow, having been formed then filled almost completely by younger flows, leaving only parts of the rims untouched.[*][1] Some also have landslide debris deposits near the base of the caldera rims, and several have since been breached (their sides have collapsed). Small cones lay along the seafloor surrounding the main formation. The southeastern volcanoes also have low-level parallel faults arranged linearly against the Juan de Fuca Ridge.[*][1]

The construction of the cones left a lot of debris scattered across the ocean surface, much of which is aligned parallel to the ridge axis. The largest of these cones are nearly 2 km (1 mi) across, whereas the smallest appear to be only a few hundred meters in diameter.*[1]

In 2006, an expedition was conducted by the Monterey Bay Aquarium Research Institute on the seamount formations.*[2]

4.57.1   See also

- List of submarine volcanoes

- President Jackson Seamounts

4.57.2   References

[1] "Mapping Program: Vance Seamounts". Monterey Bay Aquarium Research Institute. Feb 6, 2009. Retrieved 28 December 2009.

[2] "Cruise into the Classroom: Vance Expedition". Monterey Bay Aquarium Research Institute. July 24 – August 6, 2006. Retrieved 28 December 2009.

4.57.3   External links

- 2006 expedition

4.58   Winslow Reef, Phoenix Islands

Main article: Kiribati

Winslow Reef is an underwater feature of the Phoenix Islands, Republic of Kiribati, located 200 km north-northwest of McKean Island at 01°36′S 174°57′W / 1.600°S 174.950°W. It is the northernmost and westernmost feature of the Phoenix Islands, not counting the outlying Baker and Howland Islands. It has a least depth of 11 metres. The reef is about 1.6 km long east-west, and about half that wide. The bottom is pink coral and red sand. Winslow Reef is mentioned by Robert Louis Stevenson, who sailed over an area thought to be Winslow Reef in late 1889, but did not find it.*[1]

The reef was discovered by the whaler *Phoenix* in 1851, and the name of the whaler became attached to the entire group of islands.*[2] Perry Winslow was the master of the *Phoenix* on this voyage.*[3]

It is part of the Phoenix Islands Protected Area and is therefore a protected nature reserve.

The Winslow Reef borders the U.S. Howland-Baker EEZ.*[4] The PacIOOS mentions that Winslow Reef is "on the southeast boundary line of the EEZ".*[5]

It is not to be confused with Winslow Reef in the Cook Islands at 20°38′S 160°56′W / 20.633°S 160.933°W.

4.58.1   See also

- List of Guano Island claims

- Carondelet Reef

4.58.2 References

[1] Stevenson, Robert Louis (August 1998). "Chapter 13, Part V". *Letters of Robert Louis Stevenson* (e-book). Volume 2. Seattle, Washington, USA.: The World Wide SchoolTM. Retrieved 2008-12-17. We had one particularity: coming down on Winslow Reef, p. d. (position doubtful): two positions in the directory, a third (if you cared to count that) on the chart; heavy sea running, and the night due. The boats were cleared, bread put on board, and we made up our packets for a boat voyage of four or five hundred miles, and turned in, expectant of a crash. Needless to say it did not come, and no doubt we were far to leeward.

[2] Denger, Otto; Gillaspy, Edwin (August 15, 1955). *Atoll Research Bulletin, Canton Island, South Pacific* (PDF) **41**. Washington DC: Pacific Science Board, National Academy of Sciences-National Research Council. p. 6. Retrieved 2008-12-17. The whaler 'Phoenix' discovered Winslow Reef, northwest of Canton, in 1851, and the name of this vessel became attached to the entire group of islands.

[3] Ships' Log Collection, *Phoenix*, Nov. 7, 1848 – Feb. 5, 1853. In the Nantucket Historical Association, Resource Library and Archives.

[4] "Phoenix Island protected area. Management plan, 2009-2014." (pdf). UNESCO. Retrieved 2012-02-20.

[5] "PacIOOS. Howland & Baker". Pacific Islands Ocean Observing System. Retrieved 2012-02-23.

4.58.3 External links

- Phoenix Islands Protected Area, Kiribati

- Seamount Catalog —Winslow Reef Seamount listing

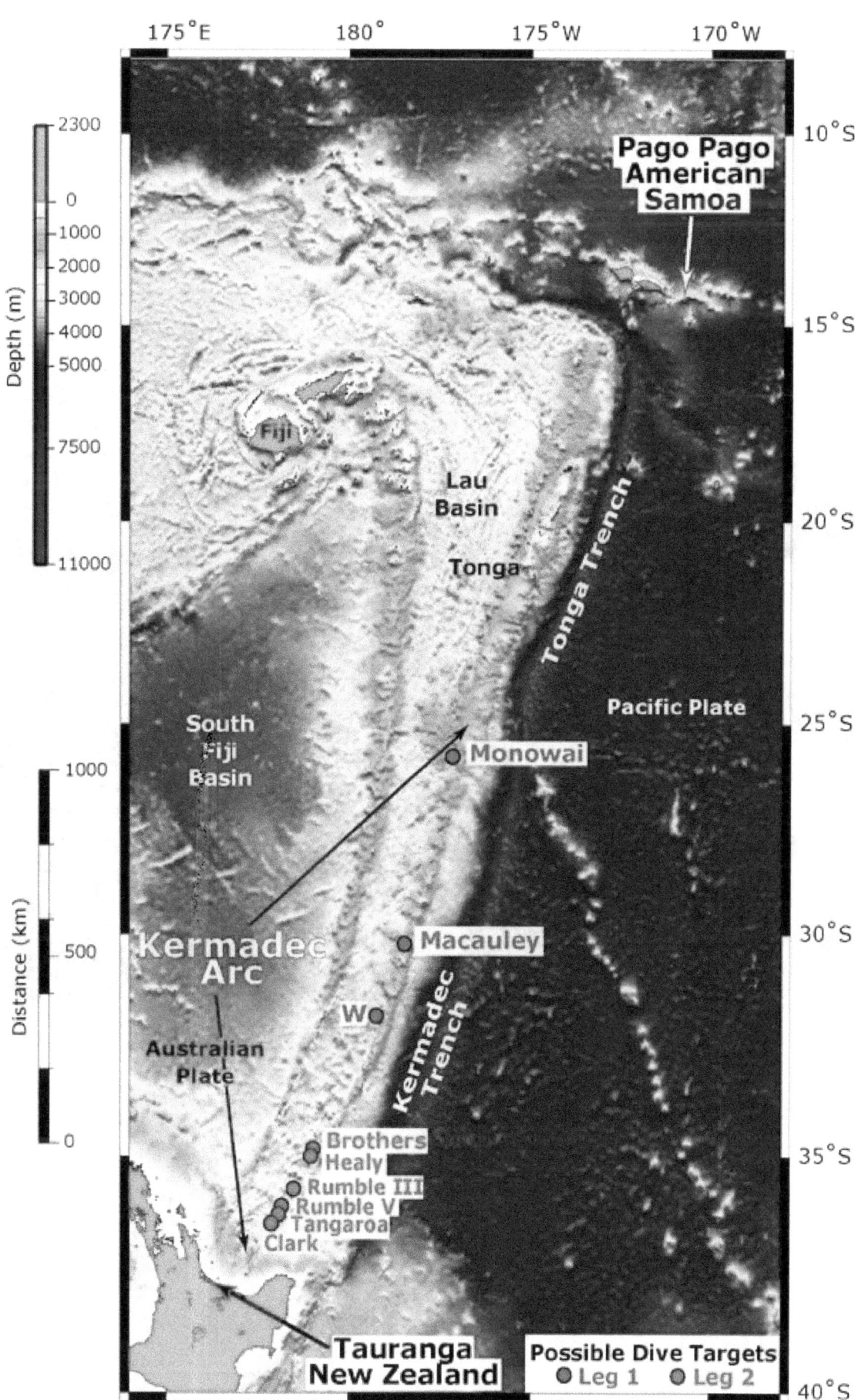

Bathymetric map of Kermadec islands and seamounts

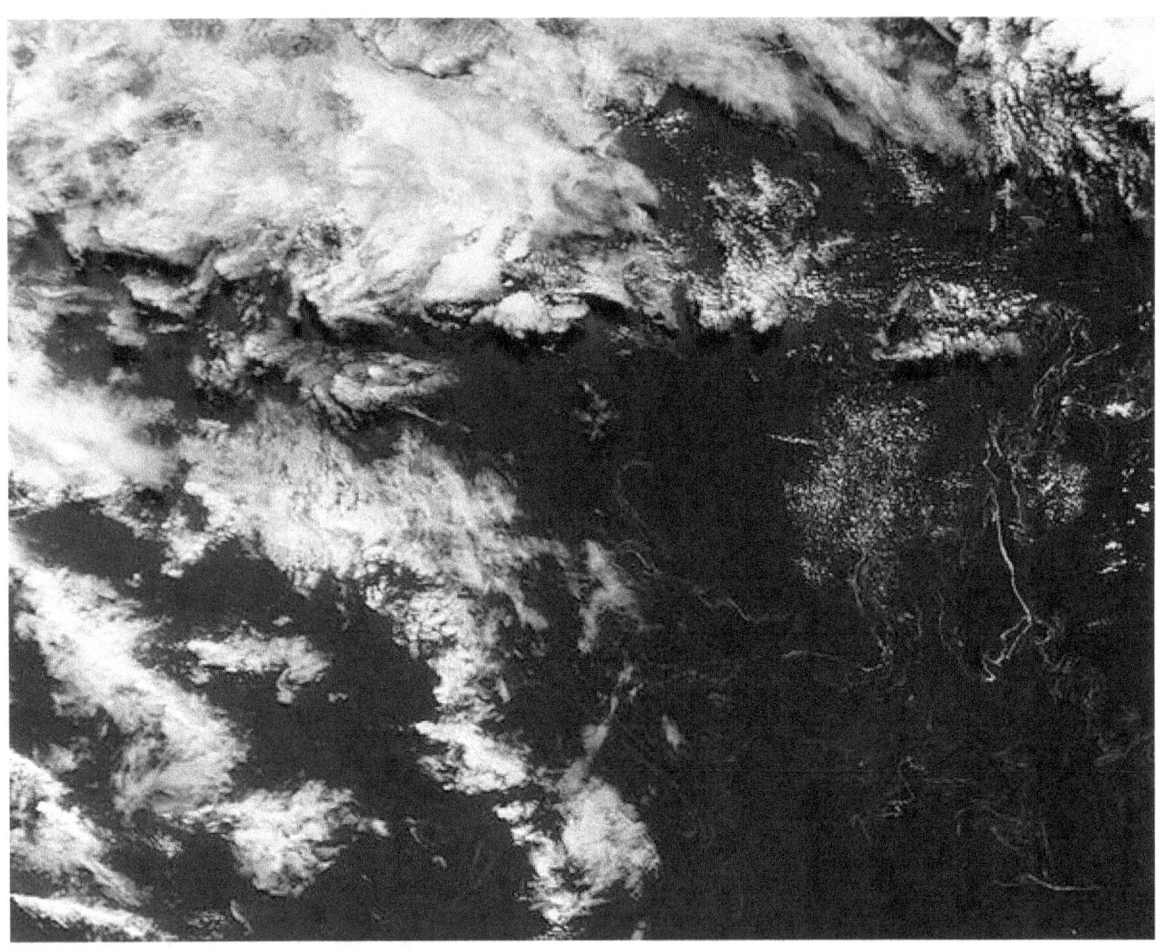

Kermadec Islands pumice raft on 12 August 2012. Raft can be seen as fibrous tendrils primarily in lower right quadrant; Raoul Island can be seen as green dot near upper right.

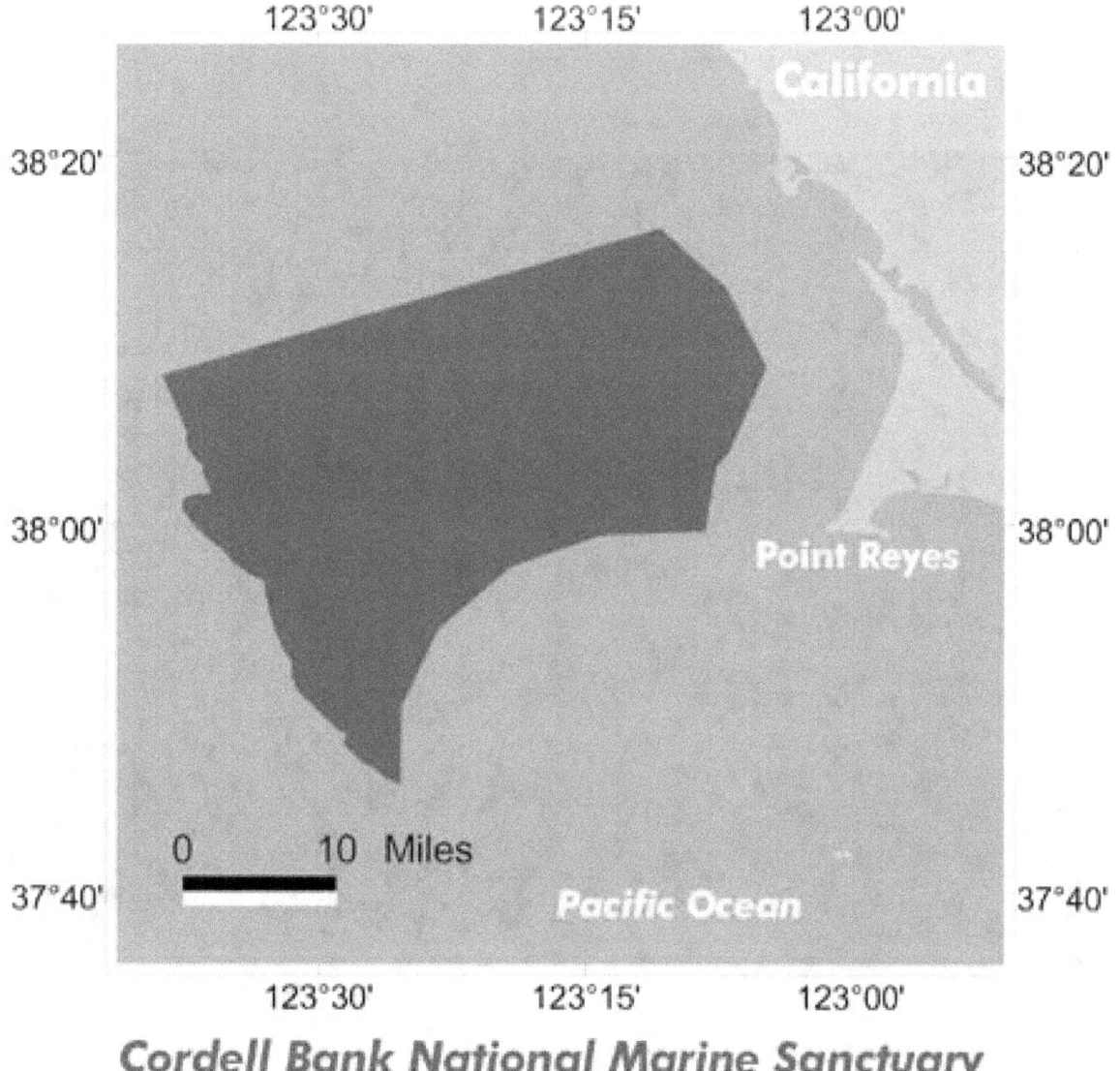

Cordell Bank National Marine Sanctuary

Area map of the sanctuary The coast of California is to the upper right, and the administrative centre, located on the granite outcropping, adjacent to it. The actual sanctuary is further left, and colored darker blue.

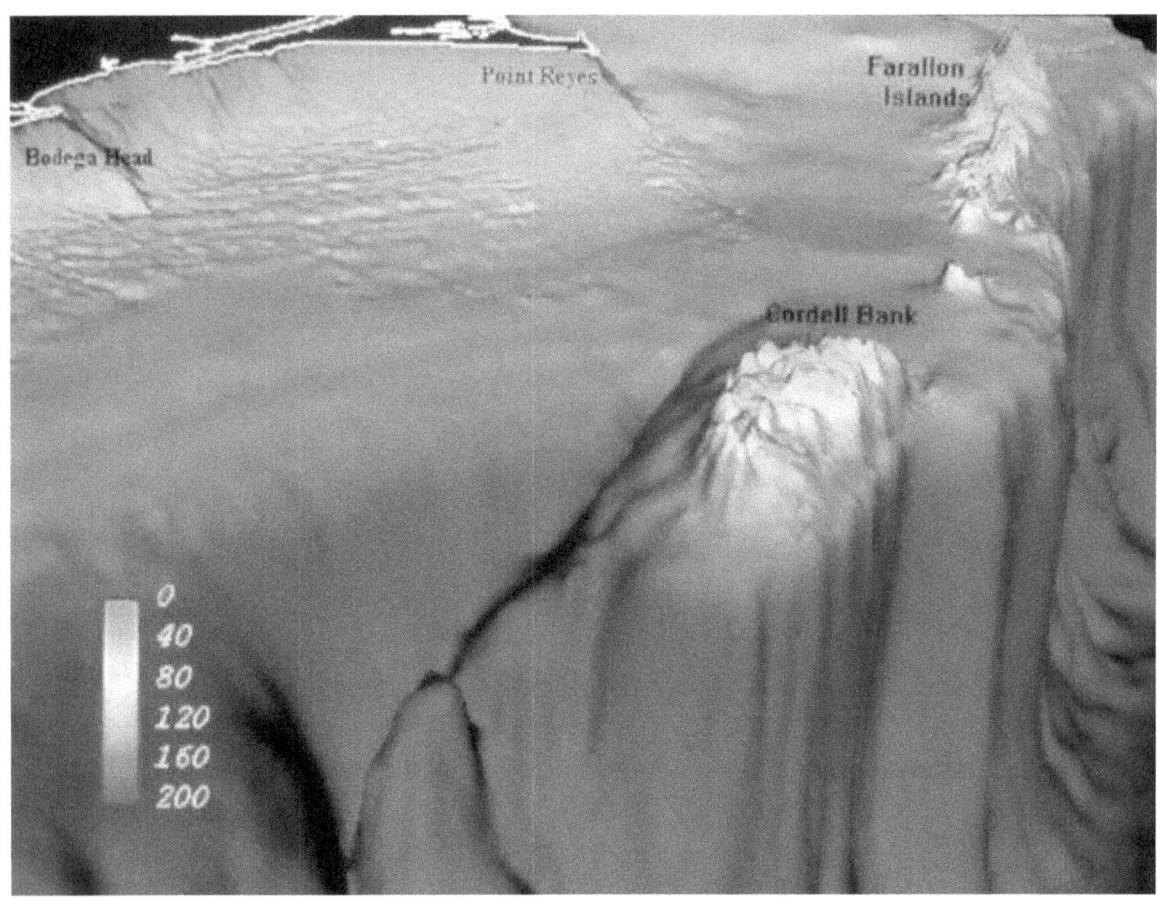

Underwater topography of Cordell Bank showing seamount and nearby Farallon Islands

Cordell Bank - Rosy Rockfish and Strawberry Anemones at 55.5 meters depth

Peter Mel at Cortes Bank

Bigeye tuna, one of Cross's two common fish species.

George Davidson, for whom the seamount was named.

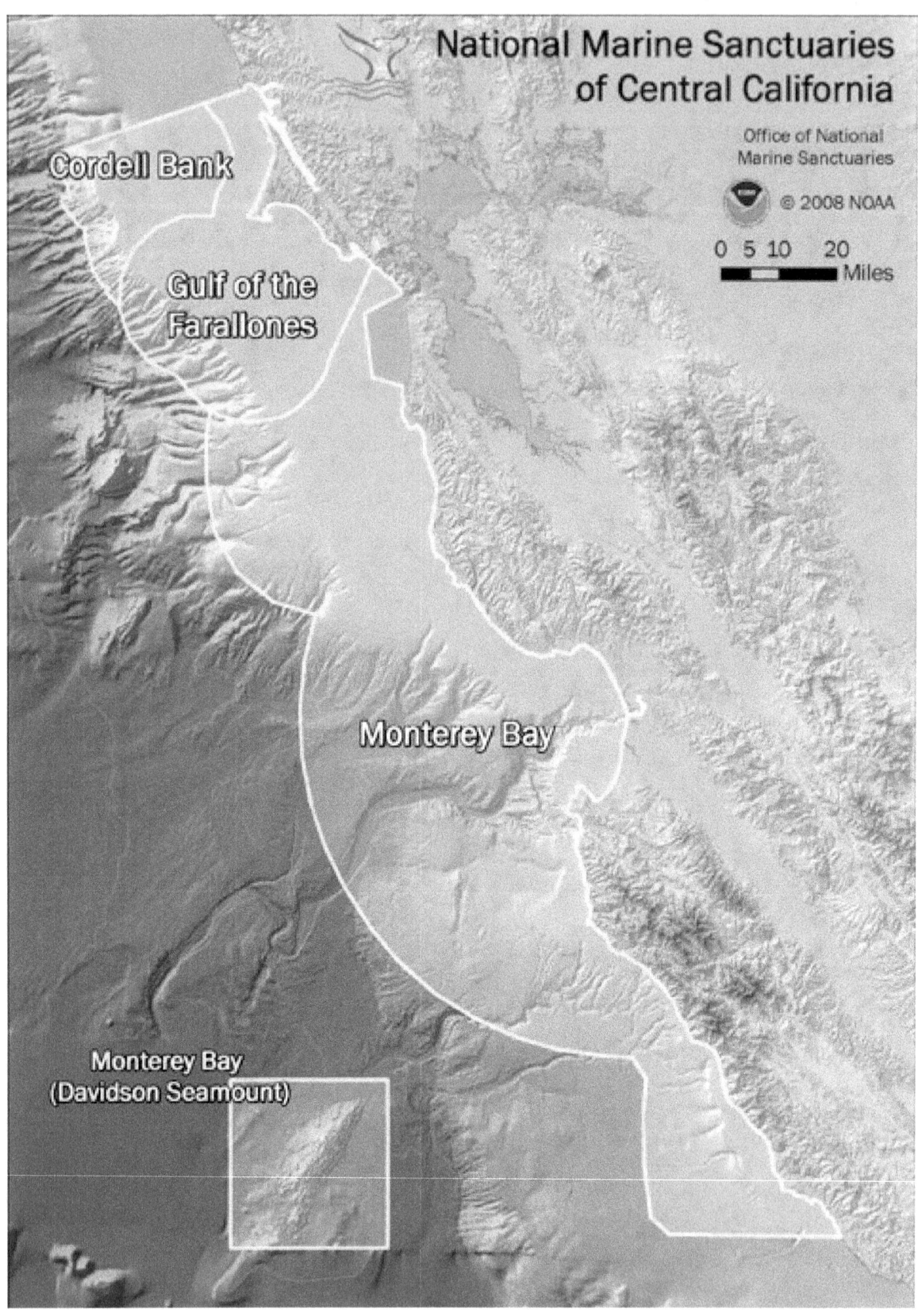

Diagram illustrating the orientation of the 3 marine sanctuaries of Central California: Cordell Bank, Gulf of the Farallones, and Monterey Bay. Davidson Seamount, part of the Monterey Bay sanctuary, is indicated at bottom-left.

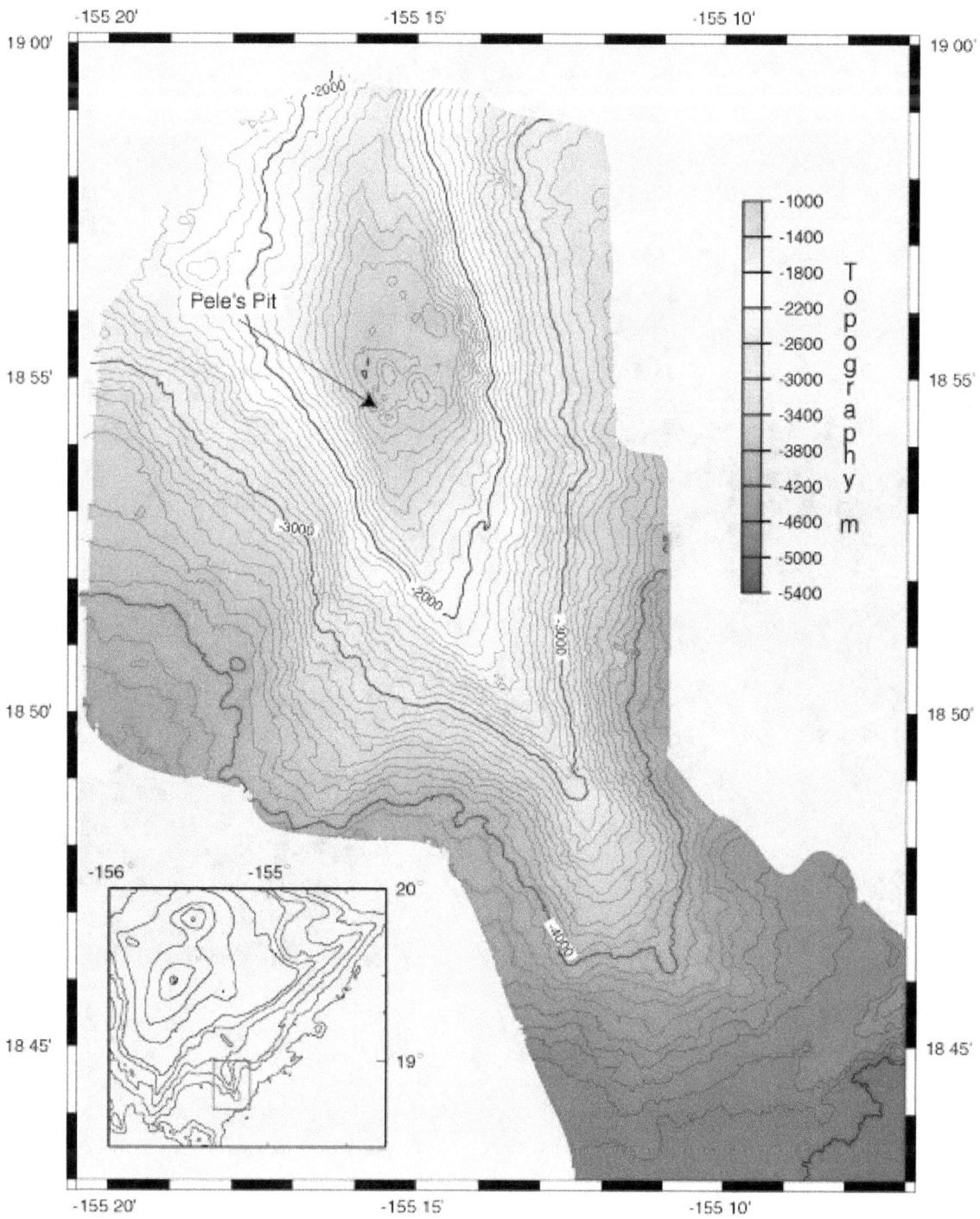

Bathymetric mapping of Lōʻihi; the arrow points to Pele's Pit.

Depth (m)

988

2035

3085

4130

5178

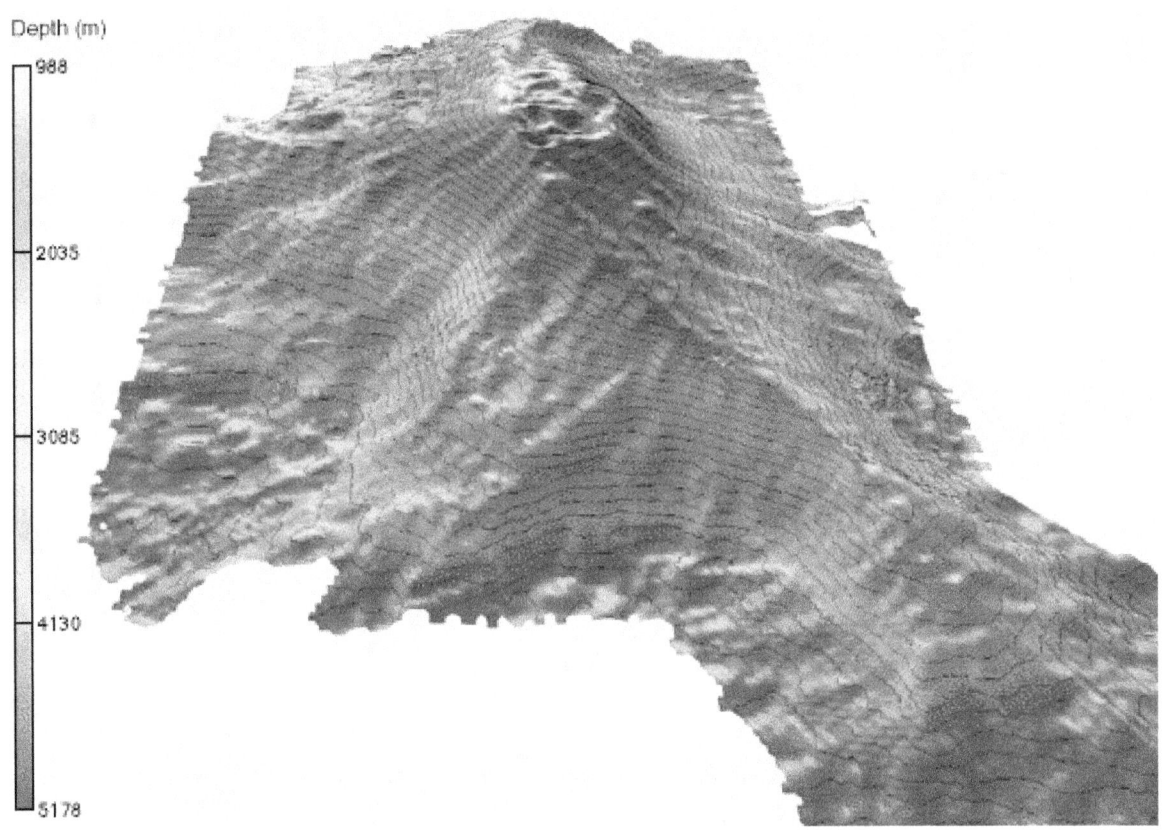

Three dimensional rendering of the Seamount

R/V (research vessel) Kaʻimikai-o-Kanaloa (KoK) launching Pisces V, a battery-powered submersible. The R/V KoK is the support ship for the Hawaiʻi Undersea Research Laboratory (HURL).

Ocean bottom observatory (OBO) at Pele's Vents

Exposed pillow lava in the Northern range.

A specimen of Solenosmilia variabeles, *a species of coral that forms the base of Tasmanian Seamounts' ecosystem*

Chapter 5

Seamounts of the Southern Ocean, Mediterranean & Indian Ocean

5.1 Adare Seamounts

Adare Seamounts, also known as **Adare Mountains**, are the seamounts in Balleny Basin named in association with Adare Peninsula and Cape Adare. Name approved by the Advisory Committee on Undersea Features, June 1988.

Coordinates: 70°0′S 171°30′E / 70.000°S 171.500°E This article incorporates public domain material from the United States Geological Survey document "Adare Seamounts" (content from the Geographic Names Information System).

5.2 Balleny Seamounts

Balleny Seamounts (61°0′S 161°30′E / 61.000°S 161.500°ECoordinates: 61°0′S 161°30′E / 61.000°S 161.500°E) are seamounts named in association with the Balleny Islands. The name was approved by the Advisory Committee for Undersea Features in June 1988.

5.2.1 References

- ⊘ This article incorporates public domain material from the United States Geological Survey document "Balleny Seamounts" (content from the Geographic Names Information System).

5.3 Barsukov Seamount

Barsukov Seamount (61°3′S 29°12′W / 61.050°S 29.200°WCoordinates: 61°3′S 29°12′W / 61.050°S 29.200°W) is a seamount named in honor of the Russian scientist, Valeri Barsukov, former Director of the Vernadsky Institute of Geochemistry. The name proposed by Dr. G.B. Udintsev, of the same institute, and approved by the Advisory Committee for Undersea Features in June 1995.

5.3.1 References

- This article incorporates public domain material from the United States Geological Survey document "Barsukov Seamount" (content from the Geographic Names Information System).

5.4 Belgica Guyot

Belgica Guyot (65°30′S 90°30′W / 65.500°S 90.500°WCoordinates: 65°30′S 90°30′W / 65.500°S 90.500°W) is an undersea tablemount named for the Belgian research ship *Belgica*, used in the first Belgian Antarctic Expedition (1896–1899). The name was proposed by Dr. Rick Hagen of the Alfred Wegener Institute for Polar and Marine Research, Bremerhaven, Germany, and approved by the Advisory Committee for Undersea Features in June, 1997.

5.4.1 References

- ⊘ This article incorporates public domain material from the United States Geological Survey document "Belgica Guyot" (content from the Geographic Names Information System).

5.5 Boomerang Seamount

The **Boomerang Seamount** is an active submarine volcano, located 18 km northeast of Amsterdam Island, France. It was formed by the St. Paul hotspot and has a 2 km wide caldera that is 200 m deep. Hydrothermal activity occurs within the caldera.[*][1][*][2]

5.5.1 See also

- List of volcanoes in French Southern and Antarctic Lands

5.5.2 References

[1] Boomerang Seamount: the active expression of the Amsterdam-St. Paul hotspot, Southeast Indian Ridge, K.T.M. Johnson, D.W. Graham, K.H. Rubin, K. Nicolaysen, D.S. Scheirer, D.W. Forsyth, E.T. Baker, L.M. Douglas-Priebe, Earth and Planetary Science Letters 183 (2000) 245-259

[2] "Boomerang Seamount". *Global Volcanism Program*. Smithsonian Institution.

5.6 Christmas Island Seamount Province

The **Christmas Island Seamount Province** (also known as the **Christmas Island Seamounts**) is an unusual seamount (submarine volcano) formation named for Christmas Island, an Australian territory and wildlife reserve that is also part of the chain. The province consists of more than 50 seamounts, up to 4,500 m (14,800 ft) in height, within a 1,080,000 km^2 (417,000 sq mi) area.[*][1][*][2][*][3]

Unlike most seamount groups, the Christmas Island seamount formation does not form a long hotspot-based chain of increasingly older volcanoes, instead being a scattered grouping of volcanoes within a large radius. The origins of the formation have long been enigmatic for scientists; the Christmas Island area does not exhibit the hotspot chain formation that most seamount groups have, nor does it run perpendicular to a local rift zone, instead lying roughly parallel to the edge of the Australian Plate. Many of the seamounts are flat-topped guyots, showing that at one point the province was likely a group of active volcanic islands, before it was slowly eroded to its current subsurface level.[*][2][*][3]

A study, published in 2011 and led by Kaj Hoernle of the University of Kiel in Germany, acquired and tested rock samples for ^{40}Ar/^{39}Ar, strontium, neodymium, hafnium and lead to determine its age and basis. The study found that the seamounts' rock was more similar to continental than oceanic crust, particularly resembling northwest Australian crust. The seamounts were found to be 47 to 136 million years old, decreasing in age from east to west, and at most 25 million years younger than the crust surrounding them. Plate reconstructions based on these dates showed that the seamounts formed where West Burma separated from Australia and India, during the breakup of Gondwana, approximately 150 million years ago. Hoernle and his associates proposed that the seamounts are made of recycled, delaminated continental crust enriched in mantle that was rising beneath the mid-ocean ridge forming at the time, and that this may be a relatively common process in shallow-basin areas.*[1]*[2]*[3]

5.6.1 References

[1] Crystal Gammon (20 December 2011). "Surprising Christmas Island Seamounts Mystery Solved". LiveScience through Yahoo News. Retrieved 30 December 2011.

[2] K. Hoernle, F. Hauff, R. Werner, P. van den Bogaard, A. D. Gibbons, S. Conrad, and R. D. Müller (27 November 2011). "Origin of Indian Ocean Seamount Province by shallow recycling of continental lithosphere". *Nature Geoscience* (Nature Publishing Group) **4**: 883–887. Bibcode:2011NatGe...4..883H. doi:10.1038/ngeo1331. Retrieved 30 December 2011.

[3] K. Hoernle; F. Hauff; R. Werner; P. Van Den Bogaard; S. Conrad; A. Gibbons; D. Muller (2009). "The Christmas Island Seamount Province, Indian Ocean: Origin of intraplate volcanism by shallow recycling of continental lithosphere?" (PDF). *Goldschmidt Conference Abstracts*.

Coordinates: 13°S 106°E / 13°S 106°E

5.7 Dallmann Seamount

Dallmann Seamount (67°10′S 96°53′W / 67.167°S 96.883°WCoordinates: 67°10′S 96°53′W / 67.167°S 96.883°W) is a seamount named for polar explorer Eduard Dallmann, who surveyed the area west of Graham Land up to 66°S. The name was proposed by Dr. Rick Hagen of the Alfred Wegener Institute for Polar and Marine Research, Bremerhaven, Germany, and was approved by the Advisory Committee for Undersea Features in June 1997.

5.7.1 References

- ⊚ This article incorporates public domain material from the United States Geological Survey document "Dallmann Seamount" (content from the Geographic Names Information System).

5.8 De Gerlache Seamounts

De Gerlache Seamounts (65°0′S 90°30′W / 65.000°S 90.500°WCoordinates: 65°0′S 90°30′W / 65.000°S 90.500°W) are seamounts in Antarctica, named for Lieutenant Adrien Victor Joseph de Gerlache (Royal Belgian Navy), Commander of the *Belgica* during the first Belgian Antarctic Expedition, 1896–1899.

5.8.1 References

- This article incorporates public domain material from the United States Geological Survey document "De Gerlache Seamounts" (content from the Geographic Names Information System).

5.9 Empedocles (volcano)

Empedocles is a large underwater volcano located 40 km off the southern coast of Sicily named after the Greek philosopher Empedocles who believed that everything on Earth was made up of the four elements.

According to Giovanni Lanzafame, of Italy's National Institute of Geophysics and Volcanology, the volcanic structure is around 400 meters high, with a base 30 km long and 25 km wide. Located in the Campi Flegrei del Mar di Sicilia

(Phlegraean Fields of the Strait of Sicily), Empedocles is composed of what was once believed to be separate volcanic centers, including Graham Island.[*][2][*][3]

The volcano shows no sign of erupting in the near future. While the volcano's top is now 7 meters below sea level, it was once visible above the water. In 1831 Empedocles broke the surface as Fernandinea Island and almost caused a major international incident when several countries tried to claim ownership of it. It disappeared into the water again five months later.[*][4]

5.9.1 References

[1] Coinage article describing island significance First paragraph, second line. Accessed February 11th, 2009

[2] Volcano larger than Washington D.C. discovered, CNN

[3] Scientists discover huge underwater volcano, The Independent Online

[4] Massive underwater volcano found, CBBC Newsround (published and accessed on June 23, 2006).

Coordinates: 37°09′49″N 12°43′07″E / 37.16361°N 12.71861°E

5.10 Eratosthenes Seamount

The **Eratosthenes Seamount** is a seamount in the Eastern Mediterranean about 100 km south of western Cyprus. It is a large, submerged massif, about 120 km long and 80 km wide. Its peak lies at the depth of 690 m and it rises 2000 m above the surrounding seafloor, which is located at the depth of up to 2,700 m and is a part of the Eratosthenes Abyssal Plain. It is one of the largest features on the Eastern Mediterranean seafloor.

5.10.1 See also

- CenSeam

- Ferdinandea

5.10.2 References

- Mart, Yossi and Robertson, Alastair H. F. (1998). *Eratosthenes Seamount: an oceanographic yardstick recording the Late Mesozoic-Tertiary geological history of the Eastern Mediterranean*, in Robertson, A.H.F., Emeis, K.-C., Richter, C., and Camerlenghi, A. (eds.), Proceedings of the Ocean Drilling Program, Scientific Results, Vol. 160, Chapter 52, 701–708.

- Kempler, Ditza (1998). *Eratosthenes Seamount: the possible spearhead of incipient continental collision in the Eastern Mediterranean*, in Robertson, A.H.F., Emeis, K.-C., Richter, C., and Camerlenghi, A. (eds.), Proceedings of the Ocean Drilling Program, Scientific Results, Vol. 160, Chapter 53, 709–721.

- Earthref entry

5.11 Graham Island (Sicily)

Graham Island (Graham Bank or Graham Shoal), also known as **Isola Ferdinandea** in Italian, is a submerged volcanic island discovered when it last appeared on August 1, 1831, by Humphrey Senhouse, the captain of the Royal Navy flagship St Vincent and named after Sir James Graham, the First Lord of the Admiralty. It was claimed by the United Kingdom. It forms part of the underwater volcano Empedocles, 30 km (19 mi) south of Sicily, and which is one of a number of submarine volcanoes known as the Campi Flegrei del Mar di Sicilia. Currently a seamount, eruptions have raised it above sea level several times before erosion submerged it again. When it last rose above sea level after erupting in 1831, a four-way dispute over its sovereignty began, which was still unresolved when it disappeared beneath the waves again in early 1832. During its brief life, the French geologist Constant Prévost was on hand, accompanied by an artist, to witness it in July 1831; he named it *Île Julia*, for its July appearance, and reported in the *Bulletin de la Société Géologique de France*.[5][6] Some observers at the time wondered if a chain of mountains would spring up, linking Sicily to Tunisia and thus upsetting the geopolitics of the region.[2] More recently, it has shown signs of volcanic activity in 2000 and 2002, forecasting a possible appearance; however, as of 2014 it remains 6 m (20 ft) under sea level.

5.11.1   Early history

Graham Island lies in a volcanic area known as the *Campi Flegrei del Mar di Sicilia* (Phlegraean Fields of the Sea of Sicily), in between Sicily and Tunisia in the Mediterranean Sea. Many submarine volcanoes (seamounts) exist in the region, as well as some volcanic islands such as Pantelleria. Volcanic activity at Graham Island was first reported during the First Punic War, and the island has appeared and disappeared four or five times.[7] Several eruptions have been reported since the 17th century.[8]

5.11.2   1831 eruption and British possession

Graham Island (Sicily), guarded by HMS Melville.

Graham Island's most recent appearance as an island was in July 1831. The first sign of an eruption was a period of high seismic activity spanning from June 28 to July 10 reported by the nearby town of Sciacca.[1] On July 4 an odor of sulfur spread through the town reportedly in such quantities that it blackened silver.[1] On July 13, a column of smoke was clearly seen from St. Domenico. The residents believed it to be a ferry on fire.[1] On the same day, the brig *Gustavo* passed through the area, confirming a bubbling in the sea that the captain thought was a sea monster. Another ship reported dead fish floating in the water. By July 17, a fully grown islet had formed.[1]

On August 1, 1831 Humphrey Fleming Senhouse, the captain of the Royal Navy flagship HMS *St Vincent* claimed the Island for the British Crown and named it after Sir James Graham, the First Lord of the Admiralty. The eruptions of 1831 resulted in the island increasing in size to about 4 km (2.5 mi). However, it was composed of loose tephra, easily eroded by wave action, and when the eruptive episode ended it rapidly subsided, disappearing beneath the waves in January 1832, before the issue of its sovereignty could be resolved. Fresh eruptions in 1863 caused the island to reappear briefly before again sinking below sea level. At its maximum (in July and August 1831), it was 4,800 m (15,700 ft) in circumference and 63 m (207 ft) in height.*[1] It sported two small lakes, the larger of which was 20 metres (66 ft) in circumference and 2 m (6 ft 7 in) in depth.*[1]

5.11.3 Dispute

Graham Island was subject to a four-way dispute over its sovereignty, originally claimed for the United Kingdom and given the name Graham Island. The King of the Two Sicilies, Ferdinand II, after whom Sicilians named the island Ferdinandea, sent ships to the nascent island to claim it for the Bourbon crown. The French Navy also made a landing, and called the island Julia. Spain also declared its territorial ambitions.*[2] Each wanted the island for its useful position in the Mediterranean trade route (to England and France) and its close position to Spain and Italy.*[2]

Initial conflict

In August 1831 the volcano had risen to above sea level, although still only a couple of rocks, but the Royal Navy thought it was very suitable as a base to control the traffic in the Mediterranean, as it was closer to the European continent than the island of Malta. The small volcanic point was an important strategic point in the Mediterranean to the world's premier maritime power of the time, being closer to Spain and Italy than Malta, the next closest.*[1] The British fleet landed, named it Graham Island, after Sir James Graham, the First Lord of the Admiralty, and planted their flag, the Union Jack.*[3]

But the King of Sicily also realized its strategic significance, and dispatched the corvette *Etna* to claim the new land and dub it Ferdinandea in honor of King Ferdinand II. Last on the scene was Constant Prévost, a co-founder of the French Geological Society, who compared the eruption to a bottle of champagne being uncorked. He named the island Julia, because it was born in July. Diplomatic wrangling broke out.*[3]

Extended conflict

For five months conflict raged in newspapers and elsewhere as the different nations fought over a roughly 60 m (200 ft) high piece of basalt.*[9] Tourists traveled to the island to see its two small lakes. Sailors watched it when passing by, and nobles of the House of Bourbon reportedly planned to set up a holiday resort on its beaches. None of these ideas came to fruition, however, as the island soon sank back beneath the waters. By December 17, 1831, officials reported no trace of it. As dynamically as the seamount appeared, it disappeared, defusing the conflict with it.*[10]

5.11.4 Recent activity

After 1863 the volcano lay dormant for many decades, its summit just 8 m (26 ft) below sea level. In 2000, renewed seismic activity around Graham Island led volcanologists to speculate that a new eruptive episode could be imminent, and the seamount might once again become an island.*[4] To forestall a renewal of the sovereignty disputes, Italian divers planted a flag on the top of the volcano in advance of its expected resurfacing.*[10] To bolster their case, Sicilians, who call it Ferdinandea, summoned the descendant of the Bourbon King of Naples. In a ceremony filmed by a flotilla of camera crews, Prince Carlo di Bourbon lowered a plaque into the waves and told cheering locals: "It will always be Sicilian." Lobbied by fishermen and sailors, Ignazio Cucchiara, the mayor of Sciacca, invited Prince Carlo to attend the ceremony with his wife, The Duchess of Castro Camilla Crociani. To accommodate television crews the plaque was lowered well before reaching the shoal, which is a danger to shipping. Choppy waters forced divers to postpone the operation a week, until November 13, 2000.*[4] The diving crew planted Sicily's flag, which features a Medusa's head surrounded by three naked legs – a sign traditionally interpreted as "keep away."

The marble plaque, weighing 150 kg (330 lb), was inscribed "This piece of land, once Ferdinandea, belonged and shall always belong to the Sicilian people." *[2] The Prince told cheering locals: "It will always be Sicilian." But within six months it had been fractured into 12 pieces, mostly likely by fishing gear but possibly by vandalism.*[2]*[6]

In November 2002, Professor Enzo Boschi, from the Institute of Geophysics and Volcanology in Rome, told BBC News Online:

> We have observed minor seismic activity, gas emissions but this is quite normal.*[10]

He put the time of resurfacing at a couple of weeks or months. However, in an interview with *Time* magazine, Boris Behncke, a German researcher at the University of Catania's department of geological sciences in Sicily, said:

> Geologically speaking, it's a possibility, But geology has a very long time scale ... We really should not be too worried.*[3]

Despite showing signs in both 2000 and 2002, the seismicity did not lead to volcanic eruptions and as of 2006 Ferdinandea's summit remains about 6 metres (20 ft) below sea level. Should it reappear, Federico Eichberg, an international relations expert based in Rome, believes it would do so within Italian territorial waters —and in all probability would be formally claimed by Italy. Eichberg does not expect that a renewed international rumpus would arise, noting:

> "If it's just a little island, we're not going to have a big fight over it." *[3]

5.11.5   Significance

Scientific study

The sudden geologic phenomenon was observed and studied by numerous scientists. Among the Germans were Hoffmann, Schultz, and Philippi. Among the English were Edward Davy and Warington Wilkinson Smyth. Among the French was Constant Prévost. Among the Italians there was Scinà Domenico (1765-1837) who published his observations in the "Effeméridi Sicilians" (1832- Vol. 2), and Carlo Gemmellaro (1787–1866), teacher of geology and mineralogy at Catania University, who published "Actions of the Gioenia Academy of Catania" (1831- Vol.8).*[1]

In 2006, further study revealed Graham Island to be just one part of the larger volcanic cone Empedocles.*[7]

Marine significance

Graham Island is still referenced on marine charts, as its top is only 6 meters short of breaking the surface, much higher than the draft of most seafaring vessels.*[9] It is also a small shoal on which near-surface maritime creatures dwell.

Coinage

In Sicily in 2000 there was produced an unofficial minting of a penny, featuring the island of Graham Island on one side and, unusually, a bust of Elizabeth II on the other. (Italy, including Sicily, was using the italian lira by this time and the coin did not circulate.) The designer of the coin is David Mannucci. The idea to make this coin occurred to Mannucci after he "found out the existence of the ghost island" from a newspaper article. Besides the copper piece, varieties exist in silver, copper "with protective enamel", and in silver "with protective enamel". While this Italian-made coin fittingly bears the Italian name for the island, the conflicted piece also features a bust of "Elizabeth II D.G.R." and bears a British denomination.*[2]

In popular culture

During its emergence it was visited by Sir Walter Scott, and it provided inspiration for James Fenimore Cooper's *The Crater, or Vulcan's Peak*, Alexandre Dumas, père's *The Speronara* and Jules Verne's *Captain Antifer* and *The Survivors of the Chancellor*. It also provided the inspiration for the isle of Leshp in Terry Pratchett's *Jingo*.

5.11.6   See also

- List of volcanoes in Italy
- Volcanism in Italy

5.11.7   References

[1] Siclian Almanac Accessed on February 11th, 2009

[2] Coinage Database Entry. Accessed February 11th, 2009

[3] Maryann Bird (March 20, 2000). "Fire from The Sea". *Time Magazine*. Retrieved 2011-10-01.

[4] http://www.guardian.co.uk/international/story/0,,396512,00.html *Bourbons surface to retake island* - Guardian Unlimited

[5] "Notes sur l' ile Julia pour servir a l' histoire de la formation des montagnes volcaniques" in *Mémoires de la Soc. Géol. de France*, 1835 ("L' exploration de île Julia")

[6] From out the azure main, Media monitor, London Geological Society, February 2003. Accessed February 20th, 2009.

[7] Scientists discover huge underwater volcano, The Independent Online

[8] *Campi Flegrei Mar Sicilia*, Smithsonian Global Volcanism Program, accessed 9 May 2006

[9] Volcanoes and Volcanism Entry. Accessed February 10, 2009

[10] BBC News Story "The last time the island surfaced, diplomatic arguments arose over its ownership;" Direct Quote accessed on February 10th, 2009

5.12   Hakurei Seamount

Hakurei Seamount (62°52′S 140°49′E / 62.867°S 140.817°ECoordinates: 62°52′S 140°49′E / 62.867°S 140.817°E) is a seamount located off Adélie Land, Antarctica. The name, approved by the Advisory Committee for Undersea Features in July 1999, is for the RV *Hakurei-maru* which conducted a detailed survey of the area.[*][1]

5.12.1   References

[1] "Hakurei Seamount". *Geographic Names Information System*. United States Geological Survey. Retrieved 2012-05-15.

This article incorporates public domain material from the United States Geological Survey document "Hakurei Seamount" (content from the Geographic Names Information System).

5.13   Iselin Seamount

Iselin Seamount is a seamount in the southern ocean off Antarctica. It was named for the research ship *Iselin II* of the Woods Hole Oceanographic Institute, the name being approved by the Advisory Committee for Undersea Features in February 1964.[*][1]

5.13.1   References

[1] "Iselin Seamount". *Geographic Names Information System*. United States Geological Survey. Retrieved 2012-07-10.

This article incorporates public domain material from the United States Geological Survey document "Iselin Seamount" (content from the Geographic Names Information System). Coordinates: 70°45′S 178°15′W / 70.750°S 178.250°W

5.14   Lecointe Guyot

Lecointe Guyot (65°6′S 93°0′W / 65.100°S 93.000°WCoordinates: 65°6′S 93°0′W / 65.100°S 93.000°W) is an undersea tablemount named for Georges Lecointe, navigator/astronomer aboard the Belgica. The name was proposed by Dr. Rick Hagen of the Alfred Wegener Institute for Polar and Marine Research, Bremerhaven, Germany, and was approved by the Advisory Committee for Undersea Features in June 1997.[*][1]

5.14.1   References

[1] "Lecointe Guyot". *Geographic Names Information System*. United States Geological Survey. Retrieved 2013-06-07.

This article incorporates public domain material from the United States Geological Survey document "Lecointe Guyot" (content from the Geographic Names Information System).

5.15   Lichtner Seamount

Lichtner Seamount (67°33′S 0°40′W / 67.550°S 0.667°WCoordinates: 67°33′S 0°40′W / 67.550°S 0.667°W) is a seamount located in the Southern Ocean. The name, for German cartographer Werner Lichtner, was approved by the Advisory Committee for Undersea Features in April 2000.[*][1]

5.15.1   References

[1] "Lichtner Seamount". *Geographic Names Information System*. United States Geological Survey. Retrieved 2013-06-14.

This article incorporates public domain material from the United States Geological Survey document "Lichtner Seamount" (content from the Geographic Names Information System).

5.16   Marsili

For the surname, see Marsili (surname).

Marsili is a large undersea volcano in the Tyrrhenian Sea, about 175 kilometers (109 mi) south of Naples. The seamount is about 3,000 m (9,800 feet) tall; its peak and crater are about 450 m below the sea surface. Though it has not erupted in recorded history, volcanologists believe that Marsili is a relatively fragile-walled structure, made of low-density and unstable rocks,[*][1] fed by the underlying shallow magma chamber. Volcanologists with the Italian National Institute of Geophysics and Volcanology (INGV) announced on March 29, 2010 that Marsili could erupt at any time, and might experience a catastrophic collapse that would suddenly release vast amounts of magma in an undersea eruption and landslide that could trigger destructive tsunamis on the Italian coast and nearby Mediterranean coastlines.[*][2]

5.16.1   See also

- List of volcanoes in Italy

- Volcanism in Italy

- Seamount

- Submarine volcanoes

5.16.2 References

[1] Caratori Tontini F., Cocchi L., Muccini F., Carmisciano C., Marani M., Bonatti E., Ligi M., and Boschi E., Potential-field modelling of collapse-prone submarine volcanoes in the Southern Tyrrhenian Sea (Italy), *Geophysical Research Letter* **37** (2010), L03305, doi:10.1029/2009GL041757.

[2] "Undersea volcano threatens southern Italy: report". AFP. March 29, 2010. Retrieved August 16, 2014.

5.17 Maud Seamount

Maud Seamount (65°0′S 2°35′E / 65.000°S 2.583°ECoordinates: 65°0′S 2°35′E / 65.000°S 2.583°E) is a seamount in the Southern Ocean. Its name was approved by the Advisory Committee for Undersea Features in February 1964.[*][1]

5.17.1 References

[1] "Maud Seamount". *Geographic Names Information System*. United States Geological Survey. Retrieved 2013-08-29.

This article incorporates public domain material from the United States Geological Survey document "Maud Seamount" (content from the Geographic Names Information System).

5.18 Muirfield Seamount

The **Muirfield Seamount** is a submarine mountain located in the Indian Ocean approximately 130 kilometres (70 nautical miles) southwest of the Cocos (Keeling) Islands. The Cocos Islands are an Australian territory, and therefore the Muirfield Seamount is within Australia's Exclusive Economic Zone (EEZ). The Muirfield Seamount is a submerged archipelago, approximately 2.5 kilometres (1.6 miles) in diameter and 16–18 metres (52–59 feet) below the surface of the sea. A 1999 biological survey of the seamount performed by the Australian Commonwealth Scientific and Industrial Research Organisation (CSIRO) revealed that the area is depauperate.

The Muirfield Seamount was discovered accidentally in 1973 when the cargo ship *MV Muirfield* (a merchant vessel named after Muirfield, Scotland) was motoring in waters charted at a depth of greater than 5,000 metres (16,404 ft), when she suddenly struck an unknown object, resulting in extensive damage to her keel.[*][1] In 1983, HMAS *Moresby*, a Royal Australian Navy survey ship, surveyed the area where *Muirfield* was damaged, and charted in detail this previously unsuspected hazard to navigation.

The dramatic accidental discovery of the Muirfield Seamount is often cited as an example of limitations in the vertical datum accuracy of some offshore areas as represented on nautical charts, especially on small-scale charts. More recently, in 2005 the submarine USS San Francisco (SSN-711) ran into an uncharted seamount about 560 kilometers (350 statute miles) south of Guam at a speed of 35 knots (40.3 mph; 64.8 km/h), sustaining serious damage and killing one seaman.

5.18.1 See also

- Jasper Seamount

- Graveyard Seamounts

- Mud volcano

- Sedlo Seamount

- South Chamorro Seamount

5.18.2   References

[1] Calder, Nigel. *How to Read a Navigational Chart: A Complete Guide to the Symbols, Abbreviations, and Data Displayed on Nautical Charts*. International Marine/Ragged Mountain Press, 2002.

5.18.3   External links

- "Franklin Voyage Summary No. FR07/99". CSIRO. 2003.

5.19   Orca Seamount

Orca Seamount is a seamount (underwater volcano) near King George Island and Cape Sherriff Field Station in Antarctica, in the Bransfield Strait. It is inactive.

The crater rim is about 3 km wide and about 500 m above the ocean floor.[*][1]

The seamount was first named by Professor O. González-Ferrán of Chile in 1987, after the orca (killer whale) often sighted in these waters.[*][2][*][3] It was mapped and studied by the ship RV *Polarstern* during an Antarctic cruise (number ANT-XI/3) in 2005.[*][4] The variant name of **Viehoff Seamount** (approved in 6/95 ACUF 263) was named for Dr. Thomas Viehoff, a remote sensing specialist in marine sciences. Name proposed by Dr. G.B. Udintsev, Vernadsky Institute of Geochemistry (VIG).[*][5]

5.19.1   References

[1] Hatzky, Jörn (2005): The Orca Seamount Region, Antarctica (Sect. 5.5.2). In: Peter C. Wille (ed.), Sound Images of the Ocean in Research and Monitoring, Springer-Verlag Berlin.

[2] "Name Details – Orca Seamount". *Catalogue of Antarctic Names*. Australian Antarctic Data Centre. Retrieved 2010-03-29.

[3] "GEBCO Gazetteer of Geographic Names of Undersea Features" (PDF). GEBCO Sub-Committee on Undersea Feature Names (SCUFN). January 2010. p. 332. Retrieved 2010-03-29.

[4] *Vulcano bransfield-strait hg.png*

[5] "Data Query". *Geographic Names Information System*. USGS. Retrieved 2009-05-12.

5.20   Rosenthal Seamount

Rosenthal Seamount (68°38′S 97°5′W / 68.633°S 97.083°WCoordinates: 68°38′S 97°5′W / 68.633°S 97.083°W) is a seamount in the Weddell Sea named for Alfred Rosenthal (1828–1882), German captain and ship owner involved in polar research. Rosenthal financed Dallmann's voyage. Name proposed by Dr. Heinrich Hinze, Alfred Wegener Institute for Polar and Marine Research, Bremerhaven, Germany. Name approved 6/97 (ACUF 271).

This article incorporates public domain material from the United States Geological Survey document "Rosenthal Seamount" (content from the Geographic Names Information System).

5.21   Walters Shoals

The **Walters Shoals** is a group of submerged mountains off the coast of Madagascar. Despite being 700 kilometres (430 mi) from the coast, the tips of some of the mountains are only 50 metres (160 ft) below the surface. The Walters Shoals is home to many species of fish, crustaceans, and mollusks. It was discovered in 1963 by the South African Hydrographic Frigate SAS *Natal* captained by Cmdr Walters. When found it had a huge population of Galápagos sharks but they have since been fished out.

5.21.1 References

5.22 Wordie Seamount

Wordie Seamount is a seamount located in Bransfield Strait, Antarctica. The feature is named after James Wordie, geologist on Ernest Shackleton's 1914 expedition to Antarctica.[1][2]

5.22.1 Location

Wordie Seamount is located at 61°48′S 55°27′W / 61.800°S 55.450°WCoordinates: 61°48′S 55°27′W / 61.800°S 55.450°W, which is 37 km south of Gibbs Island in the South Shetland Islands.

5.22.2 References

[1] Wordie Seamount. SCAR Composite Antarctic Gazetteer.

[2] GEBCO Gazetteer of Geographic Names of Undersea Features.

Ferdinand II of the Two Sicilies, after whom the island was named, circa 1850.

A seismograph, placed on the underwater island in 2006, is recovered in 2007.

A page out of the field journal of French geologist Constant Prévost. Illustrations by a French artist.

French geologist Constant Prévost

Italian geologist Carlo Gemmellaro (it)

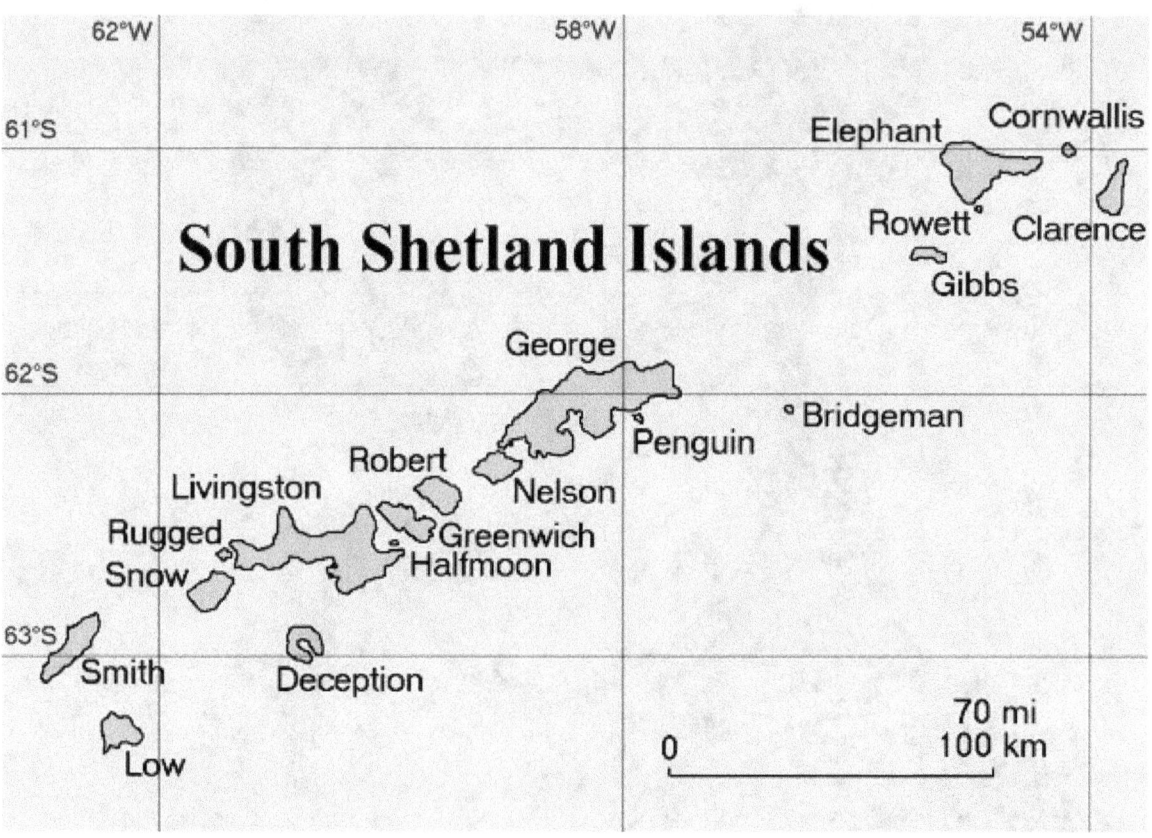

South Shetland Islands

Chapter 6

Seamount chains

6.1 Rio Grande Rise

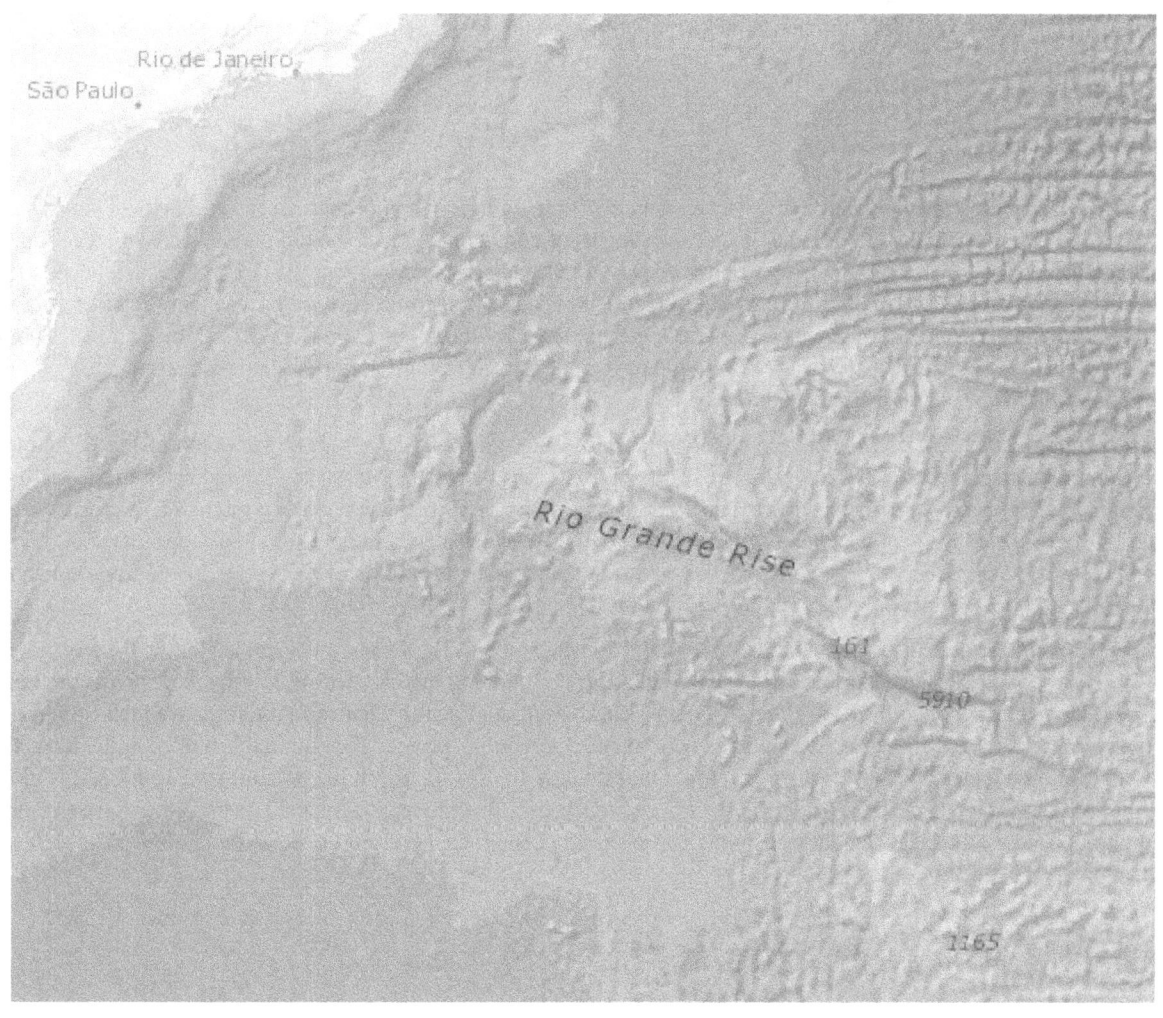

The Rio Grande Rice separates the Brazil (north) and Argentine Basins (south) and is separated from the Vema Sill and Santos Plain (west) by the Vema Channel and from the Mid-Atlantic Ridge by the Hunter Channel (east).[1]

The **Rio Grande Rise** is an aseismic ocean ridge in the southern Atlantic Ocean off the coast of Brazil. Together with the Walvis Ridge off Africa, the Rio Grande Rise forms a V-shaped structure of mirrored hotspot tracks or seamount chains across the northern South Atlantic.[2] In 2013, Brazilian scientists announced that they found granite boulders on the Rio Grande Rise and speculated that it could be the remains of a submerged continent, which they called the "Brazilian Atlantis". Other researchers, however, noted that such boulders can end-up on the ocean floor by less speculative means.[3]

6.1.1 Geology

The Rio Grande Rise separates the Santos and Pelotas Basins and is composed of western and eastern areas, which have different geological backgrounds. The western area has numerous guyots and seamounts and a basement dated to 80 to 87 million years ago. The eastern area is covered by fracture zones and may represent an abandoned spreading centre. In the western area, volcanic breccia and layers of ash indicate widespread volcanism during the Eocene, which coincides with the formation of volcanic rocks onshore. During this period, parts of the western plateau were uplifted over sea level and short-lived volcanic islands formed.[4]

When West Gondwana (i.e. South America) broke away from Africa during the Early Cretaceous (146 to 100 Ma), the South Atlantic opened up from its southern to its northern end. In this process, the voluminous Paraná and Etendeka continental flood basalts formed in what is now Brazil and Namibia. This event is linked to the Tristan-Gough hotspot, now located near the Mid-Atlantic Ridge, close to Tristan da Cunha and the Gough Islands. During the Maastrichtian (60 Ma), the orientation of spreading changed, which is still visible on the African side, and volcanism ended on the American side. This process resulted in the Tristan-Gough seamount chains on either side of the Tristan-Gough hotspot.[4]

Palaeoclimatic role

A Brazilian-Japanese expedition in 2013 recovered *in situ* granitic and metamorphic rocks on the Rio Grande Rise. This can possibly indicate that the plateau includes fragments of continental crust —possible remains of microcontinents similar to those found on and around Kerguelen in the Indian Ocean and Jan Mayen in the Arctic Ocean. The existence of such microcontinents is speculative, however, since their remains tend to be covered by younger layers of lava and sediments.[4] Nevertheless, transoceanic dispersals are hinted at by the fossil record of, for example, flightless birds such as *Lavocatavis*, indicating that several islands between Africa and South America made island hopping possible across the Atlantic during the Tertiary (66 to 2.58 Ma).[5]

At the beginning of the Maastrichian, the characteristics of water masses differed north and south of the Rio Grande Rice-Walvis Ridge complex. The disappearance of these differences during the Maastritchian indicates a reorganisation of oceanic circulation patterns that lead to a global homogenisation of intermediate and deep waters. This process seems to have been triggered by the breaching of the Rio Grande Rise-Walvis Ridge complex and the disappearance of epicontinental seaways such as the Tethys Ocean. The process resulted in the deterioration of rudist-dominated tropical habitats and consequently the extinction of benthic inoceramid bivalves.[6]

The origin of modern circulation of cold, deep water —known as the "Big Flush" —is associated with Early Eocene (55 to 40 Ma) geological events; tectonism that resulted in the opening of the north-east Atlantic and fracture zones that developed in the subsiding Rio Grande Rise, which allowed cold water from the Antarctic Weddell Sea to flow northward into the North Atlantic. 40 Ma, the generation of cold bottom water in the Antarctic resulted in the formation of psychrospheric fauna, which today live in temperatures below 10 °C (50 °F), in the Atlantic and Tethys. This global distribution suggests that the Rio Grande Rise had been breached by this time, allowing cold, dense water to move north-south through a corridor enhancing the transition from a latitudinal thermospheric circulation to a meridional thermohaline circulation.[7]

6.1.2 References

Notes

[1] Zenk & Morozov 2007, Fig. 1

[2] O'Connor & Duncan 1990, Introduction, p. 17475

[3] National Geographic News 2013

[4] Sager 2014, pp. 2–4

[5] Mourer-Chauviré et al. 2011, Abstract

[6] Frank & Arthur 1999, Conclusions, p.115

[7] Berggren 1982, Cenozoic, pp. 122-123

Sources

- Berggren, W. A. (1982). "Role of ocean gateways in climatic change" . In Geophysics Committee, Berger, W. H.; Crowell, J. C. *Climate in Earth History* (PDF, 25 Mb). Studies in Geophysics. Washington D.C.: National Academy Press. pp. 118–125. Retrieved May 2015.

- Frank, T. D.; Arthur, M. A. (1999). "Tectonic forcings of Maastrichtian ocean-climate evolution" (PDF). *Paleoceanography* **14** (2): 103–117. doi:10.1029/1998PA900017. Retrieved May 2015.

- Mourer-Chauviré, C.; Tabuce, R.; Mahboubi, M.; Adaci, M.; Bensalah, M. (2011). "A Phororhacoid bird from the Eocene of Africa" . *Naturwissenschaften* **98** (10): 815–823. doi:10.1007/s00114-011-0829-5.

- O'Connor, J. M.; Duncan, R. A. (1990). "Evolution of the Walvis Ridge-Rio Grande Rise Hot Spot System: Implications for African and South American Plate motions over plumes" (PDF). *Journal of Geophysical Research: Solid Earth (1978–2012)* **95** (B11): 17475–17502. Retrieved May 2015.

- Sager, W. W. (2014). "Scientific Drilling in the South Atlantic: Rio Grande Rise, Walvis Ridge and surrounding areas" (PDF). *U.S. Science Support Program Workshop Report*. Retrieved May 2015.

- Than, Ken (May 11, 2013). "Lost Land Found by Scientists" . National Geographic News. Retrieved June 2015.

- Zenk, W.; Morozov, E. (2007). "Decadal warming of the coldest Antarctic Bottom Water flow through the Vema Channel" . *Geophys. Res. Lett.* **34** (14): L14607. doi:10.1029/2007GL030340.

Coordinates: 31°S 35°W / 31°S 35°W

6.2 Walvis Ridge

The **Walvis Ridge** (*walvis* means whale in Dutch and Afrikaans) is an aseismic ocean ridge in the southern Atlantic Ocean. More than 3,000 km (1,900 mi) in length, it extends from the Mid-Atlantic Ridge, near Tristan da Cunha and the Gough Islands, to the African coast (at 18°S).[1] The Walvis Ridge is one of few examples of a hotspot seamount chain that links a flood basalt province to an active hotspot. It is also considered one of the most important hotspot tracks because the Tristan Hotspot is one of few primary or deep mantle hotspots.[2]

6.2.1 Geology

Apart from the Mid-Atlantic Ridge, the Walvis Ridge and the Rio Grande Rise are the most distinctive feature of the South Atlantic sea floor. They originated from hotspot volcanism and together they form a mirrored symmetry across the Mid-Atlantic Ridge, with the Tristan Hotspot at its centre. Two of the distinct sections in the Walvis Ridge have similar mirrored regions in the Rio Grande Rise; for example, the eastern section of the Walvis Ridge evolved in conjunction with the Torres Arch (the western end of the Rio Grande Rise, off the Brazilian coast) and, as the South Atlantic gradually opened, these structures became separated. The complex of seamounts in the western end of the Walvis Ridge, however, does not have a similar structure on the American side, but there is a Zapiola Seamount Complex south of the eastern end

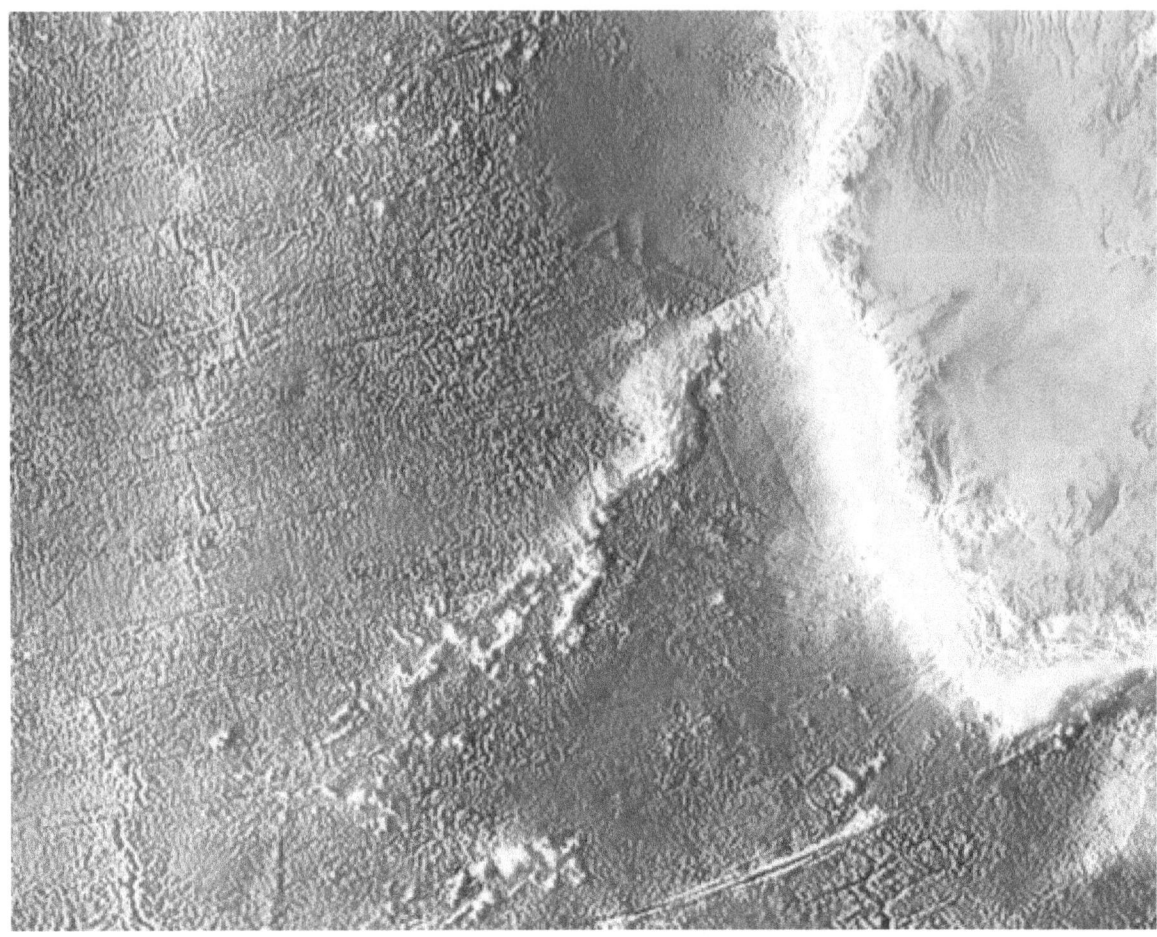

The Walvis Ridge stretches some 3,000 km (1,900 mi) from the African continental shelf to the Tristan da Cunha hotspot, separating the Angola and Cape Basins.

of the Rio Grande Rise.[*][3] The formation of this mirrored structure is the result of the opening of the South Atlantic some 120 Mya and the Paraná and Etendeka continental flood basalts, the lateral-most parts of the structure, formed at the beginning of this process in areas that are now located in Brazil and Namibia.[*][2]

The Walvis Ridge is divided into three main sections:[*][1]

1. A first 600 km (370 mi) long segment stretching from Africa to approximately longitude 6°E and varying in width between 90–200 km (56–124 mi).

2. A second section, 500 km (310 mi) long, stretching north-south, and narrower than the first section.

3. A third more discontinuous section, which is marked by seamounts and connects the Walvis Ridge to the Mid-Atlantic Ridge.

Cretaceous kimberlites in the central Democratic Republic of Congo and Angola align with the Walvis Ridge.[*][4]

The Tristan-Gough hotspot track first formed over the mantle plume that formed the Etendeka-Paraná continental flood basalts some 135 to 132 Ma.[*][5] The eastern section of the ridge is thought to have been created in the Middle Cretaceous period, between 120 to 80 Ma.[*][6][*][7] While the mantle plume remained large and stable, the eastern Walvis Ridge formed along with the Rio Grande Rise over the Mid-Atlantic Ridge.[*][5] During the Maastrichtian 60 million years ago, the orientation of spreading changed, which is still visible in the orientation of the various sections of the Walvis Ridge.[*][2] The mantle plume then gradually became unstable and bifurcated 60 to 70 Ma to produce the two separate

Tristan and Gough hotspot tracks. It finally disintegrated 35 to 45 Ma and formed the guyot province in the western end of the ridge.[*][5]

Hundreds of volcanic explosions were recorded on the Walvis Ridge in 2001 and 2002. These explosions seemed to come from an unnamed seamount on the northern side of the ridge and are thought to be unrelated to the Tristan hotspot.[*][8]

The Ewing Seamount is part of the ridge.

Palaeoclimatic role

The Eocene Layer of Mysterious Origin (Elmo) is a period of global warming that occurred 53.7 Ma, about two million years after the Paleocene–Eocene Thermal Maximum. This period manifests as a carbonate-poor red clay layer unique to the Walvis Ridge and is similar to the PETM, but of smaller magnitude.[*][9][*][10]

6.2.2 Oceanography

The Walvis Ridge is a natural obstacle for the Agulhas rings, mesoscale warm core rings that are shed from the Agulhas Current south of the Agulhas Bank. In average, five such rings are shed each year, a number that varies considerably between years.[*][11] The rings tend cross the Walvis Ridge at its deepest part, but they still lose transitional speed and many rings decay rapidly.[*][12] Their transitional speed drop from 5.2±3.6 km/day to 4.6±3.1 km/day, but it is not clear how much the Walvis Ridge is responsible for this drop, since the rings' speed drop to 4.3±2.2 km/day between the Walvis Ridge and the Mid-Atlantic Ridge.[*][13] The rings can cross the South Atlantic in 2.5–3 years but only two thirds make it farther than the Walvis Ridge.[*][11] When the rings pass over the Cap Basin south of the Walvis Ridge they are frequently disturbed by the Benguela Current, interaction between rings, and bottom topography such as the Vema Seamount, but there are fewer obstacles and disturbances west of the Walvis Ridge were the rings tend stabilise. [*][14] The Agulhas rings transport an estimated 1-5 Sv (millions m^2/s) of water from the Indian Ocean to the South Atlantic.[*][15]

Originating around Antarctica, Antarctic Bottom Water (AABW) enters the Cape Basin between the Agulhas Bank and the Agulhas Ridge after which it flows west north of the Agulhas Ridge. AABW then retroflects at the south-western end of the Walvis Ridge, flows north-east along the ridge before being retroflected south by North Atlantic Deep Water, with which it exits the Cape Basin and flows into the Indian Ocean.[*][16]

6.2.3 References

Notes

[1] Goslin et al. 1974, Introduction, p. 469

[2] Sager 2014, pp. 2–5

[3] O'Connor & Duncan 1990, Introduction, p. 17475

[4] de Wit 2007, Fig. 7, p. 380; Fig. 9, p. 385

[5] Rohde et al. 2013, Conclusions, pp. 69-70

[6] Pastouret & Goslin 1974

[7] Müller, Royer & Lawver 1993

[8] Haxel & Dziak 2005, Abstract

[9] Lourens et al. 2005, Abstract

[10] "Eocene Layer of Mysterious Origin". JOIDES Resolution. Retrieved May 2015.

[11] Schouten et al. 2000, Discussion and Conclusions, p. 21933

[12] Schouten et al. 2000, Abstract, Introduction, pp. 21913-21914

[13] Schouten et al. 2000, Rings paths, pp. 21916-21918

[14] Schouten et al. 2000, Ring Decay, pp. 21918-21919

[15] Ruijter et al. 2003, p. 46

[16] Gruetzner & Uenzelmann-Neben 2014, Fig 1.A

Sources

- de Wit, M. (2007). "The Kalahari Epeirogeny and climate change: differentiating cause and effect from core to space" (PDF). *South African Journal of Geology* **110** (2-3): 367–392. doi:10.2113/gssajg.110.2-3.367. Retrieved June 2015.

- Goslin, J.; Mascle, J.; Sibuet, J.; Hoskins, H. (1974). "Geophysical Study of the Easternmost Walvis Ridge, South Atlantic: Morphology and Shallow Structure" (PDF). *Geological Society of America Bulletin* **85** (4): 619–632. Retrieved May 2015.

- Gruetzner, J.; Uenzelmann-Neben, G. (2014). "Contourites at the eastern Agulhas Ridge and Cape Rise seamount shaped by Southern Ocean derived water masses" (PDF). *2nd Deep-Water Circulation Congress, 10-12 Sept. 2014, Ghent, Belgium.* Retrieved July 2015.

- Haxel, J. H.; Dziak, R. P. (2005). "Evidence of explosive seafloor volcanic activity from the Walvis Ridge, South Atlantic Ocean" (PDF). *Geophysical Research Letters* **21**: L13609. doi:10.1029/2005GL023205. Retrieved May 2015.

- Lourens, L. J.; Sluijs, A.; Kroon, D.; Zachos, J. C.; Thomas, E.; Röhl, U.; Bowles, J.; Raffi, I. (2005). "Astronomical pacing of late Palaeocene to early Eocene global warming events". *Nature* **435** (7045): 1083–1087. doi:10.1038/nature03814.

- Müller, D.; Royer, J.-Y.; Lawver, A. (1993). "Revised plate motions relative to the hotspots from combined Atlantic and Indian Ocean hotspot tracks" (PDF). *Geology* **21** (3): 275–278. Bibcode:1993Geo....21..275D.

- O'Connor, J. M.; Duncan, R. A. (1990). "Evolution of the Walvis Ridge-Rio Grande Rise Hot Spot System: Implications for African and South American Plate motions over plumes" (PDF). *Journal of Geophysical Research: Solid Earth (1978–2012)* **95** (B11): 17475–17502. Retrieved May 2015.

- Pastouret, L.; Goslin, J. (1974). "Middle Cretaceous sediments from the eastern part of Walvis Ridge". *Nature* **248** (5448): 495–496. Bibcode:1974Natur.248..495P. doi:10.1038/248495a0. Retrieved March 2009.

- Rohde, J. K.; van den Bogaard, P.; Hoernle, K.; Hauff, F.; Werner, R. (2013). "Evidence for an age progression along the Tristan-Gough volcanic track from new ^{40}Ar/^{39}Ar ages on phenocryst phases". *Tectonophysics* **604**: 60–71. doi:10.1016/j.tecto.2012.08.026. Retrieved June 2015.

- Ruijter, W. P. M., de; Cunningham, S. A.; Gordon, A. L.; Lutjeharms, J. R. E.; Matano, R. P.; Piola, A. R. (2003). "On the South Atlantic Climate Observing System (SACOS)" (PDF). *Report of the CLIVAR/OOPC/IAI workshop* (NOAA). Retrieved January 2015.

- Sager, W. W. (2014). "Scientific Drilling in the South Atlantic: Rio Grande Rise, Walvis Ridge and surrounding areas" (PDF). *U.S. Science Support Program Workshop Report.* Retrieved May 2015.

- Schouten, M. W.; Ruijter, W. P. M., de; Leeuwen, P. J., van; Lutjeharms, J. R. E. (2000). "Translation, decay and splitting of Agulhas rings in the southeastern Atlantic Ocean". *Journal of Geophysical Research: Oceans* **105** (C9): 913–921.

Coordinates: 26°S 6°E / 26°S 6°E

Chapter 7

Text and image sources, contributors, and licenses

7.1 Text

- **Seamount** *Source:* https://en.wikipedia.org/wiki/Seamount?oldid=683348948 *Contributors:* Vicki Rosenzweig, Bryan Derksen, Stevertigo, Marshman, Phoebe, Fredrik, Cholling, Auric, Saforrest, Michael Devore, Avatar, Kate, Atrian, Freakofnurture, Rich Farmbrough, Oliver Lineham, Vsmith, Florian Blaschke, Bennylin, Worldtraveller, Chris huh, Viriditas, Cavrdg, Siim, Alansohn, Buaidh, Avenue, Velella, Xover, BD2412, Jorunn, Rjwilmsi, Vegaswikian, Bbullot~enwiki, Ground Zero, Gurch, Chobot, Supasheep, Joel7687, Tjarrett, Epipelagic, Dead-EyeArrow, Mejor Los Indios, IslandHopper973, SmackBot, Jab843, Hmains, Chris the speller, Bluebot, Faint death, Gracenotes, Parent5446, Shrumster, Dogears, Gobonobo, Jon186, Peter Horn, ShakingSpirit, CmdrObot, Shikaga, Lgh, Keithh, Thijs!bot, Sselbor, Escarbot, AntiVandalBot, Joan-of-arc, Prolog, Jj137, Danger, Dgc03052, Volcanoguy, CHG, BloodDoc, MartinBot, CommonsDelinker, GeoWriter, DadaNeem, Oshwah, F.chiodo, Eve Hall, Martin451, Consalveym, SieBot, Brenont, Seuraza, Mygerardromance, Maralia, Superbeecat, ClueBot, Pimvantend, Marjaliisa, Mild Bill Hiccup, Niceguyedc, Puchiko, The Red, ErgoSum88, Monfornot, SimonKSK, Addbot, DOI bot, Fgnievinski, Moosehadley, NjardarBot, DFS454, Baffle gab1978, Numbo3-bot, Lightbot, Zorrobot, Luckas-bot, Yobot, Backslash Forwardslash, DemocraticLuntz, Jim1138, Beemovieeragon, Citation bot, Erud, AbigailAbernathy, Resident Mario, False vacuum, DannyBoy20802, D'ohBot, OgreBot, Citation bot 1, Citation bot 4, Guanlongwucaii, Jonesey95, Fama Clamosa, Imma pirate, TGCP, EmausBot, John of Reading, WikitanvirBot, GoingBatty, ZéroBot, Bahudhara, H3llBot, Rocketrod1960, ClueBot NG, Dan Hunton, Gilderien, Widr, Helpful Pixie Bot, ?oygul, KLBot2, Bibcode Bot, BG19bot, Badon, Ninjaflab, YFdyh-bot, Dexbot, Meme567, Iforgotmylegs, Jo-Jo Eumerus, Henky Yoga, Ally1133, 15petedc, Mattymmoo, Monkbot, KasparBot, PeterTHarris, Margot und Herbert and Anonymous: 85

- **Guyot** *Source:* https://en.wikipedia.org/wiki/Guyot?oldid=662732512 *Contributors:* Mav, Edward, Salsa Shark, Itai, Gilgamesh~enwiki, Big-Haz, Vsmith, RJHall, MisterSheik, Kwamikagami, Siim, Johntex, RPIRED, AlexTiefling, Rjwilmsi, FlaBot, Dantheox, Aeusoes1, Epipelagic, DeadEyeArrow, SmackBot, Srnec, Bluebot, Veggies, Soumyasch, Trasel, Agne27, Thijs!bot, Marek69, Z10x, Escarbot, AntiVandalBot, Volcanoguy, Deflective, Husond, Thehamburglar, Magioladitis, NatureA16, Filll, TomS TDotO, Lantonov, Александр Сигачёв, Iosef, Seattle Skier, Gaianauta, TXiKiBoT, F.chiodo, Btermini, Griss vigaskald, Synthebot, NHRHS2010, Sémhur, Ar2cool, Dtvjho, Extremecircuitz, Clue-Bot, DumZiBoT, MystBot, Rachelcarson123, Addbot, Lightbot, ماني, Alexandru.demian, Luckas-bot, Yobot, Azcolvin429, Tempodivalse, JoniFili, Materialscientist, Xqbot, Resident Mario, Blahlo, Ale And Quail, EmausBot, ZéroBot, 4dedsel, ClueBot NG, Dan Hunton, Noyster, Dick turpentine, Раціональне анархіст, Smartcollegeboy, KasparBot and Anonymous: 55

- **Asphalt volcano** *Source:* https://en.wikipedia.org/wiki/Asphalt_volcano?oldid=669704532 *Contributors:* Auric, Vsmith, LunarLander, Woohookitty, GdlR, Sei Shonagon~enwiki, Ser Amantio di Nicolao, Mikenorton, GeoWriter, Ceranthor, Piledhigheranddeeper, Addbot, Lightbot, The Bushranger, Citation bot, Resident Mario, Smallman12q, Girlwithgreeneyes, Abductive, Claypoolej, ScottMHoward, Trappist the monk, Innotata, RjwilmsiBot, DASHBot, Look2See1, Bahudhara, Peckoduck, Bibcode Bot and Anonymous: 4

- **Volcano** *Source:* https://en.wikipedia.org/wiki/Volcano?oldid=686399267 *Contributors:* Vicki Rosenzweig, Mav, Bryan Derksen, Tarquin, RobLa, Malcolm Farmer, Scipius, Vignaux, Fredbauder, Yooden, Aldie, Toby Bartels, PierreAbbat, Karen Johnson, William Avery, Peterlin~enwiki, Robert Foley, Daniel C. Boyer, AdamRetchless, Graham, Heron, Olivier, Frecklefoot, Infrogmation, Michael Hardy, Liftarn, lxtd64, Sannse, Cameron Dewe, Dori, Tregoweth, Gaz~enwiki, Ahoerstemeier, Muriel Gottrop~enwiki, Samuelsen, Snoyes, Angela, Julesd, Glenn, Lmb~enwiki, Andres, Evercat, TonyClarke, Lancevortex, Mxn, Conti, Pizza Puzzle, Schneelocke, Dragons Bay, Hike395, Emperorbma, Adam Bishop, Dino, Dmsar, Fuzheado, Wik, Steinsky, Tpbradbury, Marshman, Maximus Rex, Morwen, Tempshill, SEWilco, Mowgli~enwiki, Jose Ramos, Bevo, Topbanana, Renato Caniatti~enwiki, Raul654, Nosebud, Wetman, Bcorr, Veghead, Denelson83, Donarreiskoffer, Robbot, Palnu, ChrisO~enwiki, Fredrik, Korath, TMillerCA, Chris 73, Schutz, Yrjö Kari-Koskinen, ZimZalaBim, Altenmann, Nurg, Lowellian, Sverdrup, Academic Challenger, Rursus, Texture, Litefantastic, Auric, Hadal, JesseW, Robinh, Lupo, Dmn, Adam78, Dave6, Ancheta Wis, Giftlite, Dbenbenn, DocWatson42, MPF, Elf, Mrguytodd, Nadavspi, Inter, Ævar Arnfjörð Bjarmason, Tom harrison, Lupin, Mark Richards, Spencer195, Marcika, Peruvianllama, Everyking, Maha ts, Curps, NeoJustin, Gamaliel, Aoi, Quinwound, Beardo, Gilgamesh~enwiki, Mboverload, Moogle10000, Edcolins, Jurema Oliveira, Golbez, Kandar, Wmahan, Utcursch, SoWhy, Seba~enwiki, Slowking Man, Yardcock, Pcarbonn, Quadell, Fangz, Antandrus, The Singing Badger, Beland, OverlordQ, FelineAvenger, ShakataGaNai, Jossi, CaribDigita, Heman, Rdsmith4, Oneiros, DragonflySixtyseven, Maximaximax, Bumm13, Kevin B12, Icairns, GeoGreg, Joyous!, Jcw69, Jh51681, Picapica, Deglr6328,

Random account 47, Adashiel, Randwicked, Kate, Mike Rosoft, Alkivar, Dufekin, Freakofnurture, Venu62, DanielCD, EugeneZelenko, Discospinster, Solitude, Rich Farmbrough, Guanabot, Oliver Lineham, Cacycle, Pmsyyz, Vsmith, Leandros, Xezbeth, Paul August, SpookyMulder, Mario todte, Bender235, ESkog, Closeapple, Petersam, Danny B-), Hapsiainen, Brian0918, RJHall, Leperflesh, Ben Webber, El C, Kwamikagami, Worldtraveller, Shanes, Briséis~enwiki, Art LaPella, RoyBoy, Triona, Dbalsdon, Leif, Aaronbrick, Drmagic, Bobo192, Smalljim, BrokenSegue, Viriditas, Fremsley, Cmdrjameson, NightDragon, Foobaz, Ziggurat, Man vyi, La goutte de pluie, TheProject, Darwinek, Alphax, MPerel, Mpulier, A Karley, Knucmo2, Rye1967, Jumbuck, Tra, Alansohn, Inexplicable, Anthony Appleyard, Mo0, Ryanmcdaniel, Interiot, Atlant, Babajobu, Revmachine21, Andrewpmk, Davenbelle, Primalchaos, AzaToth, Lectonar, MarkGallagher, Goldom, RoySmith, Malo, Titanium Dragon, Avenue, Bart133, Hohum, Snowolf, Smcgrother, Wtmitchell, BaronLarf, BanyanTree, ClockworkSoul, Super-Magician, Helixblue, ReyBrujo, RainbowOfLight, Sciurinæ, Kusma, Gene Nygaard, Ghirlandajo, Vadim Makarov, HenryLi, Dan100, Oleg Alexandrov, Y0u, Tariqabjotu, Alex.g, Bobrayner, JALockhart, Angr, Jeffrey O. Gustafson, Henrik, Mindmatrix, TigerShark, Natcase, Jdorje, Masterjamie, Camw, Webdinger, Yansa, PoccilScript, Rocastelo, Mike123, Plek, Computerdude33, Carcharoth, Before My Ken, WadeSimMiser, MONGO, GeorgeOrr, Eleassar777, Bbatsell, Mangojuice, Greg-nz, Sengkang, GregorB, Wayward, Gimboid13, Zpb52, GSlicer, Graham87, Magister Mathematicae, BD2412, Deadcorpse, NubKnacker, RxS, Ciroa, Sjö, Sjakkalle, Rjwilmsi, Koavf, Саша Стефановић, Erebus555, D34gl3r, Vary, Eyu100, JoshuacUK, JHMM13, Tawker, Mitul0520, Nneonneo, SeanMack, Brighterorange, GregAsche, Sango123, Cassowary, Twerbrou, DirkvdM, Dbigwood, Ravidreams, Titoxd, Lpstubbs~enwiki, FlaBot, Ian Pitchford, SchuminWeb, Ground Zero, Nowhither, Latka, Winhunter, Pumeleon, Nihiltres, Alhutch, Nivix, AI, Gparker, Celestianpower, RexNL, Gurch, AdamantlyMike, Tedder, Srleffler, Gurubrahma, Imnotminkus, Smithbrenon, Chobot, Bornhj, Bjwebb, Mhking, Bdelisle, Bgwhite, Cactus.man, Digitalme, Dj Capricorn, Gwernol, Tone, EamonnPKeane, Gap, Roboto de Ajvol, The Rambling Man, Wavelength, TexasAndroid, Huw Powell, Jtkiefer, Anonymous editor, NorCalHistory, SpuriousQ, CanadianCaesar, Stephenb, Gaius Cornelius, CambridgeBayWeather, Philopedia, Wimt, Shanel, NawlinWiki, Wiki alf, Bachrach44, Astral, Robertvan1, NickBush24, Jaxl, Terfili, Howcheng, OOZ662, Stephen e nelson, Dureo, Irishguy, Nick, Anetode, Banes, Brian Crawford, Cholmes75, E rulez, Spolloman~enwiki, Ezeu, Nick C, Tony1, Bucketsofg, Aaron Schulz, Elizabeyth, Khalid!, DeadEyeArrow, Mddake, Jpeob, Brisvegas, Jverkoey, Nick123, Max Schwarz, Wknight94, Stefan Udrea, FF2010, Schnauf, NorsemanII, Jason Schlumbohm, Zet Hikari, Theda, Jwissick, Јованчб, Errabee, E Wing, NHSavage, Tsunaminoai, Claygate, JoanneB, Alasdair, Brianlucas, LeonardoRob0t, Peter, Anclation~enwiki, ArielGold, SorryGuy, Extreme Unction, Katieh5584, Kungfuadam, Elephant Juice, DVD R W, Bibliomaniac15, Marquez~enwiki, David Wahler, Sacxpert, SpLoT, Qlorplox, A bit iffy, SmackBot, Lavintzin, Enlil Ninlil, Moeron, Reedy, KnowledgeOfSelf, CopperMurdoch, NorthernFire, Shoy, Pgk, Firekid, Bomac, Jacek Kendysz, Freekee, Davewild, Thunderboltz, Midway, TobiMcIntyre, Delldot, ProveIt, Kintetsubuffalo, Edgar181, Gaff, Moralis, Yamaguchi 先生, Gilliam, Gregjgrose, Hmains, Kdliss, Skizzik, Jackk, Saros136, Chris the speller, Bluebot, Stubblyhead, Carboxen~enwiki, Oli Filth, Skomae, Liamdaly620, SchfiftyThree, Epastore, Robth, DHN-bot~enwiki, Mkamensek, Gracenotes, Nichetas, John Reaves, Manhinli~enwiki, Rama's Arrow, Diyako, Suicidalhamster, RussellMcKenzie, WikiPedant, Dethme0w, Can't sleep, clown will eat me, RyanEberhart, Nick Levine, Aquarius Rising, Jorvik, HoodedMan, Lord of the earth, Nixeagle, Sommers, Avb, EvelinaB, TKD, Adamschneider, Addshore, Kcordina, Chcknwnm, Midnightcomm, Oli b, Phantom kiwi, Khoikhoi, Nev2, Flyguy649, Cybercobra, Nibuod, Nakon, OranL, Phil 999, Jiddisch~enwiki, James McNally, Kntrabssi, RJN, Michael-Billington, Rolinator, Dreadstar, TrogdorPolitiks, Richard001, Zawthet, Andrew c, AndyBQ, Risker, Pilotguy, Kukini, Ohconfucius, Autopilot, Abi79, SashatoBot, Nishkid64, CFLeon, Robomaeyhem, Aspern, Srikeit, Zahid Abdassabur, Molerat, Kuru, Titus III, Scientizzle, Jaffer, Pat Payne, JohnCub, Fev, Soumyasch, Sir Nicholas de Mimsy-Porpington, Linnell, Benesch, JorisvS, GVP Webmaster, Goodnightmush, Wickethewok, PseudoSudo, Chrisd87, Ckatz, Chrisch, Imdugud, MarkSutton, Erika Yurken, Slakr, Werdan7, TFNorman, SkippyNZ, Mr Stephen, Emurph, GilbertoSilvaFan, Jonny1~enwiki, Novangelis, Jose77, LaMenta3, Andreworkney, Paukrus, Iridescent, Dekaels~enwiki, Lakers, Tps, Joseph Solis in Australia, JoeBot, Tmangray, Walton One, R~enwiki, Igoldste, Tony Fox, ChadyWady, CapitalR, Momet, Civil Engineer III, Courcelles, Dpeters11, Fdp, Tawkerbot2, MarylandArtLover, Dlohcierekim, Daniel5127, Raphael1, Kjkrum, J Milburn, Jeremy Banks, Taskmaster99, Makedonia, CmdrObot, Deon, JPilborough, Bordellcantride, 2772Rev, Makeemlighter, SupaStarGirl, Picaroon, CWY2190, Lmcelhiney, GHe, Dgw, Green caterpillar, ChessMan007, Chmee2, MarsRover, Lazulilasher, Shizane, Some P. Erson, Smoove Z, CJBot, Ettang, Nauticashades, Fluence, Nbound, Lesqual, ArgentTurquoise, Steel, Gogo Dodo, Travelbird, Khatru2, JFreeman, Kimyu12, Umdunno, DangApricot, ST47, Pascal.Tesson, Tracy the astonishing, Multimedia Mike, Tawkerbot4, DumbBOT, Chrislk02, FastLizard4, Lee, Gonzo fan2007, Emmett5, Omicronpersei8, Marvolo Gaunt, Daniel Olsen, Satori Son, Thirdgrade0015, Mathpianist93, Thijs!bot, Epbr123, Ryansca, Mercury~enwiki, Akuru, Tsogo3, Kablammo, Callmarcus, Sting, Wompa99, Lukeyman, Marek69, West Brom 4ever, Ufwuct, Iceman00, MesserWoland, Josen, Dfrg.msc, NigelR, Mailseth, Dgies, Emergen, Srose, Neo lmx, Escarbot, Mentifisto, Thadius856, Archaen sax, KrakatoaKatie, Sidasta, AntiVandalBot, Yuanchosaan, Joanneduff, Majorly, Saimhe, Luna Santin, Omphacite, CodeWeasel, QuiteUnusual, Carolmooredc, EarthPerson, Sean K, Willscrlt, Mary Mark Ockerbloom, Tmopkisn, Cinnamon42, EdmundSS, Farosdaughter, Tillman, Chill doubt, Gdo01, Glennwells, Gregorof, Abc30, RajeshPandey, LéonTheCleaner, Falconleaf, AubreyEllenShomo, Kendothpro, CommonSense22, Edwardtbabinski, Canadian-Bacon, Res2216firestar, JackSparrow Ninja, Mikenorton, Volcanoguy, JAnDbot, UWDI ced, Leuko, Barek, MER-C, Seddon, Plm209, Edwin ok, Andonic, Hut 8.5, East718, Howsthatfordamage2?, MSBOT, Ihsansunny, Cynwolfe, LittleOldMe, Acroterion, I80and, Magioladitis, Howisube, Canjth, Bongwarrior, VoABot II, Wesley1992, KienNNN, Farquaadhnchmn, Matdog311291, Sheepmaster, SineWave, Jespinos, Cheesemann, Thekarm, Rich257, WODUP, Mother.earth, Avicennasis, Starsimon, Ajcounter, Sanket ar, Animum, Cgingold, Ali'i, Nposs, Allstarecho, Cpl Syx, Spellmaster, Glen, DerHexer, Irishchieftain, Electriceel, Khalid Mahmood, Michael K. Edwards, Wiking, Patstuart, Cocytus, NatureA16, Gjd001, Hdt83, MartinBot, Schmloof, Scholastica547, Arjun01, Stanley Gallon, ARC Gritt, MiltonT, Smart Arse, SPERM BUBBLE, Rettetast, Anaxial, CommonsDelinker, AlexiusHoratius, Darkrarehunt, Gunkarta, Eplack, LedgendGamer, Hui love n, RockMFR, J.delanoy, Nev1, Gotyear, Kimse, Tlim7882, EscapingLife, Fizzlehizz, Ali, Uncle Dick, Maurice Carbonaro, Gogglebum, Qweop, Jerry, OohBunnies!, Iains, Qhop, Qlop, GeoWriter, Q2op, Gzkn, Acalamari, Wrcmills, Alantex, Bot-Schafter, Katalaveno, Pleasetaketicket, McSly, Mdresser, SpigotMap, Eleanor Y, Skier Dude, Milespianoforte, Bkk59678, WebHamster, Chriswiki, Plasticup, GhostPirate, Raustin976, Bushcarrot, NewEnglandYankee, DadaNeem, SJP, Rbakker99, Malerin, Jorfer, Christopher Kraus, Jackaranga, Cometstyles, Inomyabcs, WJBscribe, Paranoid Eyes, Tiggerjay, Eamick, MoeGirl4455, Inter16, Websterman92, Jseach1, TheNewPhobia, John1208, Sgeureka, SoCalSuperEagle, Mon3ybagz, ThePointblank, CardinalDan, Idioma-bot, Lights, Gsalter, Deor, Jake Wasdin, VolkovBot, CWii, Xanucia, Seattle Skier, Pm911, Barneca, PBAJ, TXiKiBoT, Zamphuor, Eve Hall, Staplegunther, Vipinhari, GDonato, Walor, Josh.douglas, Rei-bot, RobinMarks, Mlnovaaa, Qxz, Someguy1221, Retiono Virginian, Una Smith, Avatars10, Aile Striker, DennyColt, Martin451, JhsBot, Slysplace, LeaveSleaves, Maizcul, Bentley4, Andrewrost3241981, Blacksmith, MearsMan, Madhero88, Mesh920, Volcanicska, Finngall, Rjwaker, Issac 45, Mr. TQD, Synthebot, Cojo84, Janeantoinette, Peterandgill, Falcon8765, Enviroboy, Anna512, Dsm56, AjitPD, Burntsauce, Hlk 751, Peterpanluver, MCTales, Stephen.trippis, Illumini85, Elriana, Brianga, Ceranthor, Jassy2010, Truthsaid, Gurpaalbains, Dasuitekilla, Vin789, AUSTINboston1, Nagy, Erikcarlson, Symane, GavinTing, The Random Editor, Ponyo, SieBot, Hiponie2, Calliopejen1, PlanetStar,

- **Gorringe Ridge** *Source:* https://en.wikipedia.org/wiki/Gorringe_Ridge?oldid=675029731 *Contributors:* Graeme Bartlett, Thorwald, Grutness, Jaraalbe, CIreland, Zeorymer, Dawnseeker2000, Volcanoguy, KConWiki, Squids and Chips, Hugo999, VolkovBot, Raymondwinn, Cherry1000, Paul20070, Addbot, This is Paul, Lightbot, Xufanc, Resident Mario, Cruks and Anonymous: 2

- **Gosnold Seamount** *Source:* https://en.wikipedia.org/wiki/Gosnold_Seamount?oldid=284563057 *Contributors:* Volcanoguy and Cherry1000

- **Great Meteor Seamount** *Source:* https://en.wikipedia.org/wiki/Great_Meteor_Seamount?oldid=679958186 *Contributors:* Brianhe, Vsmith, Grutness, SmackBot, Hmains, Volcanoguy, NatureA16, Paracel63, DadaNeem, Cherry1000, Pimvantend, Addbot, Resident Mario, FrescoBot, Fartherred and Anonymous: 3

- **Gregg Seamount** *Source:* https://en.wikipedia.org/wiki/Gregg_Seamount?oldid=283191069 *Contributors:* Volcanoguy

- **Hodgson Seamount** *Source:* https://en.wikipedia.org/wiki/Hodgson_Seamount?oldid=570302360 *Contributors:* Egpetersen, Volcanoguy, Cherry1000 and Anonymous: 1

- **Kelvin Seamount** *Source:* https://en.wikipedia.org/wiki/Kelvin_Seamount?oldid=420183994 *Contributors:* David Biddulph, Volcanoguy, Cherry1000, Harry~enwiki, Addbot, ClueBot NG and Anonymous: 1

- **Kiwi Seamount, Atlantic Ocean** *Source:* https://en.wikipedia.org/wiki/Kiwi_Seamount%2C_Atlantic_Ocean?oldid=384483020 *Contributors:* Dramatic, Volcanoguy and Cherry1000

- **Manning Seamount** *Source:* https://en.wikipedia.org/wiki/Manning_Seamount?oldid=284564416 *Contributors:* Volcanoguy, Cherry1000 and Addbot

- **Michael Seamount** *Source:* https://en.wikipedia.org/wiki/Michael_Seamount?oldid=509556055 *Contributors:* Volcanoguy, Cherry1000 and Tyhgrewskl

- **Muir Seamount** *Source:* https://en.wikipedia.org/wiki/Muir_Seamount?oldid=599536450 *Contributors:* Woohookitty, JLaTondre, Magioladitis, Pimvantend, Mild Bill Hiccup, Lightbot, Citation bot, Resident Mario, Jonesey95 and John of Reading

- **Mytilus Seamount** *Source:* https://en.wikipedia.org/wiki/Mytilus_Seamount?oldid=284564742 *Contributors:* Volcanoguy and Cherry1000

- **Nashville Seamount** *Source:* https://en.wikipedia.org/wiki/Nashville_Seamount?oldid=284564870 *Contributors:* Volcanoguy and Cherry1000

- **New England Seamount chain** *Source:* https://en.wikipedia.org/wiki/New_England_Seamount_chain?oldid=665317851 *Contributors:* Carib-Digita, Vsmith, I9Q79oL78KiL0QTFHgyc, TLSuda, SmackBot, Jab843, Hmains, Veggies, Ken Gallager, Volcanoguy, Tonicthebrown, Hugo999, Sémhur, Fbarw, Pimvantend, Addbot, Nelson3218, Ulric1313, Resident Mario, Chris.urs-o, Citation bot 1, Valdemarasl, ClueBot NG and Anonymous: 6

- **Newfoundland Ridge** *Source:* https://en.wikipedia.org/wiki/Newfoundland_Ridge?oldid=543223670 *Contributors:* Volcanoguy, Nyttend and Cherry1000

- **Newfoundland Seamounts** *Source:* https://en.wikipedia.org/wiki/Newfoundland_Seamounts?oldid=532354291 *Contributors:* Grutness, Hmains, Backspace, Volcanoguy, The Anomebot2, Cherry1000 and Anonymous: 1

- **Panulirus Seamount** *Source:* https://en.wikipedia.org/wiki/Panulirus_Seamount?oldid=284565966 *Contributors:* Volcanoguy and Cherry1000

- **Physalia Seamount** *Source:* https://en.wikipedia.org/wiki/Physalia_Seamount?oldid=284566126 *Contributors:* Volcanoguy and Cherry1000

- **Picket Seamount** *Source:* https://en.wikipedia.org/wiki/Picket_Seamount?oldid=284566247 *Contributors:* Volcanoguy and Cherry1000

- **Protector Shoal** *Source:* https://en.wikipedia.org/wiki/Protector_Shoal?oldid=654457173 *Contributors:* Ahoerstemeier, Gilgamesh~enwiki, Sam Hocevar, Vsmith, Grutness, Avenue, Ratzer, Rjwilmsi, Scfitch, SmackBot, Goldfishbutt, Hmains, Kharker, Droll, GVP Webmaster, CmdrObot, Thijs!bot, Kahastok, Volcanoguy, The Anomebot2, Hugo999, Seattle Skier, TXiKiBoT, Detroiterbot, Addbot, Zorrobot, Yobot, Citation bot, 4dedsel, FriarTuck1981 and Anonymous: 3

- **Rehoboth Seamount** *Source:* https://en.wikipedia.org/wiki/Rehoboth_Seamount?oldid=284566420 *Contributors:* Volcanoguy and Cherry1000

- **Retriever Seamount** *Source:* https://en.wikipedia.org/wiki/Retriever_Seamount?oldid=284566593 *Contributors:* Volcanoguy and Cherry1000

- **Rosemary Bank** *Source:* https://en.wikipedia.org/wiki/Rosemary_Bank?oldid=552174582 *Contributors:* CIreland, Movementarian, Bazonka and Parrot of Doom

- **San Pablo Seamount** *Source:* https://en.wikipedia.org/wiki/San_Pablo_Seamount?oldid=284566713 *Contributors:* Volcanoguy and Cherry1000

- **Sedlo Seamount** *Source:* https://en.wikipedia.org/wiki/Sedlo_Seamount?oldid=646941896 *Contributors:* Zeimusu, Woohookitty, Peter Horn, WolfmanSF, Yobot, Resident Mario, Jonesey95 and Anonymous: 1

- **Seewarte Seamounts** *Source:* https://en.wikipedia.org/wiki/Seewarte_Seamounts?oldid=532703649 *Contributors:* Volcanoguy and Pimvantend

- **Sheldrake Seamount** *Source:* https://en.wikipedia.org/wiki/Sheldrake_Seamount?oldid=284566861 *Contributors:* Volcanoguy and Cherry1000

- **St. Helena Seamount chain** *Source:* https://en.wikipedia.org/wiki/St._Helena_Seamount_chain?oldid=655675928 *Contributors:* Grutness, Jpbowen, Hmains, Volcanoguy, Maias, Cherry1000 and Dawynn

- **Vogel Seamount** *Source:* https://en.wikipedia.org/wiki/Vogel_Seamount?oldid=284567644 *Contributors:* Volcanoguy and Cherry1000

- **Abbott Seamount** *Source:* https://en.wikipedia.org/wiki/Abbott_Seamount?oldid=684045727 *Contributors:* Hike395, Smalljim, Volcanoguy, Look2See1 and Anonymous: 2

- **Banc Capel** *Source:* https://en.wikipedia.org/wiki/Banc_Capel?oldid=663069930 *Contributors:* Grutness, Goustien and Supersixseven1

- **Bowie Seamount** *Source:* https://en.wikipedia.org/wiki/Bowie_Seamount?oldid=627108739 *Contributors:* Dino, Wetman, Bearcat, Moondyne, Wwoods, Michael Devore, Gilgamesh~enwiki, D6, Rich Farmbrough, Grutness, Eyreland, BD2412, Jaraalbe, SmackBot, Gilliam, Salamurai, Iridescent, Myasuda, Malleus Fatuorum, Volcanoguy, MetsBot, TreasuryTag, Toddst1, Plastikspork, Franamax, Piledhigheranddeeper, Ktr101, MelonBot, Addbot, Yobot, FogPrince89, John of Reading, Ngmcd, GoingBatty, ZéroBot, H3llBot, Helpful Pixie Bot, Glacialfox, BattyBot, Dick turpentine and Anonymous: 13

- **2012 Kermadec Islands eruption** *Source:* https://en.wikipedia.org/wiki/2012_Kermadec_Islands_eruption?oldid=644726613 *Contributors:* Vsmith, IronGargoyle, Neelix, Truthanado, Goustien, FrescoBot, Stormchaser89, Tirohia, Monkbot and Anonymous: 3

- **Adams Seamount** *Source:* https://en.wikipedia.org/wiki/Adams_Seamount?oldid=594590106 *Contributors:* Wwoods, Qui1che, D6, Grutness, Avenue, Ksnow, Jaraalbe, GVP Webmaster, Myasuda, Volcanoguy, MetsBot, Rémih, Seattle Skier, Goustien, Addbot, Lightbot, Crusoe8181, EmausBot, Stormchaser89, ZéroBot and Anonymous: 3

- **Bollons Seamount** *Source:* https://en.wikipedia.org/wiki/Bollons_Seamount?oldid=663069979 *Contributors:* Bearcat, Grutness, Woohookitty, BD2412, Rjwilmsi, Addbot, Lightbot, LilHelpa, Resident Mario, Crusoe8181, ZéroBot, ClueBot NG and Anonymous: 1

- **Browns Mountain** *Source:* https://en.wikipedia.org/wiki/Browns_Mountain?oldid=585897064 *Contributors:* Maias

- **Carondelet Reef** *Source:* https://en.wikipedia.org/wiki/Carondelet_Reef?oldid=660322906 *Contributors:* D6, Rich Farmbrough, Grutness, Deror avi, Ratzer, Gadget850, Renesis, Hmains, Pustelnik, Thijs!bot, The Anomebot2, Wysiwyg-nz, Goustien, DragonBot, MacedonianBoy, Kcstover, Addbot, RevelationDirect, ZéroBot, H3llBot, CactusBot, MozzazzoM, Monkbot and Anonymous: 3

- **Chelan Seamount** *Source:* https://en.wikipedia.org/wiki/Chelan_Seamount?oldid=306739836 *Contributors:* D6, Grutness, SmackBot, Volcanoguy, Sfan00 IMG and Plastikspork

- **Cobb–Eickelberg Seamount chain** *Source:* https://en.wikipedia.org/wiki/Cobb%E2%80%93Eickelberg_Seamount_chain?oldid=551334229 *Contributors:* Grutness, Jaraalbe, Tony1, CmdrObot, Volcanoguy, Tonicthebrown, Yobot, Resident Mario and Full-date unlinking bot

- **Cordell Bank National Marine Sanctuary** *Source:* https://en.wikipedia.org/wiki/Cordell_Bank_National_Marine_Sanctuary?oldid=651156714 *Contributors:* Gentgeen, Everyking, Bkonrad, CanisRufus, Sabine's Sunbird, Avenue, Wevets, Woohookitty, Ratzer, BD2412, Kbdank71, NekoDaemon, Jaraalbe, SmackBot, Davepape, Hmains, Wizardman, CmdrObot, Missvain, Eltanin, Jllm06, Schmieder, Tacothecat, Gene Hobbs, Mercurywoodrose, Entirelybs, Lightmouse, Stepheng3, Addbot, Lightbot, Yobot, AnomieBOT, Killiondude, LilHelpa, Obersachsebot, Anna Frodesiak, Resident Mario, W Nowicki, Look2See1, Bamyers99, SaberToothedWhale, Trackteur and Anonymous: 7

- **Cortes Bank** *Source:* https://en.wikipedia.org/wiki/Cortes_Bank?oldid=674746229 *Contributors:* Brainsik, Wile E. Heresiarch, Andycjp, WikiParker, Scriberius, Schluum, BlueJaeger, Kerry Raymond, Wiki alf, Wsiegmund, Hmains, Kevin Ryde, Zapvet, Son of Somebody, Magioladitis, David Eppstein, LorenzoB, SPMenefee, Rrostrom, Shawn in Montreal, Woilorio, RHodnett, Elimegrover, Goustien, Bob1960evens, Mhockey, WayneWolfe, Aceofhearts1968, Atethnekos, KitchM, Tassedethe, Lightbot, Legobot, Tangopaso, FrescoBot, Full-date unlinking bot, Kibi78704, John of Reading, BurtAlert, Coleman63, Scbc27, ClueBot NG, Reddogsix, BattyBot, Ghostwavebook, Duckduckstop, Monkbot and Anonymous: 34

- **Cross Seamount** *Source:* https://en.wikipedia.org/wiki/Cross_Seamount?oldid=630031608 *Contributors:* Woohookitty, Tabletop, Droll, Mattisse, Magioladitis, CommonsDelinker, Bppubjr, Mild Bill Hiccup, Lightbot, Ettrig, Wrelwser43, LilHelpa, Resident Mario, Shubinator, Pamdhiga, John of Reading, Frietjes, Monkbot and Anonymous: 1

- **Davidson Seamount** *Source:* https://en.wikipedia.org/wiki/Davidson_Seamount?oldid=610049576 *Contributors:* Michael Hardy, Lancevortex, Wetman, Unfree, Grutness, Woohookitty, Mandarax, Ground Zero, TeaDrinker, Jaraalbe, Wavelength, Dysmorodrepanis~enwiki, Epipelagic, Gadget850, Speight, AnOddName, Valfontis, RomanSpa, Casliber, Mattisse, Volcanoguy, Ling.Nut, TreasuryTag, StAnselm, Svick, EoGuy, Mild Bill Hiccup, Piledhigheranddeeper, MystBot, Addbot, Lightbot, Ulric1313, Citation bot, LilHelpa, J JMesserly, Resident Mario, OgreBot, Citation bot 1, ZéroBot, ClueBot NG, Bibcode Bot and Anonymous: 9

- **Dellwood Seamounts** *Source:* https://en.wikipedia.org/wiki/Dellwood_Seamounts?oldid=655092316 *Contributors:* RedWolf, D6, Grutness, SmackBot, Volcanoguy, Sfan00 IMG, Plastikspork and Otolemur crassicaudatus

- **Denson Seamount** *Source:* https://en.wikipedia.org/wiki/Denson_Seamount?oldid=541801825 *Contributors:* GreatWhiteNortherner, SmackBot, Hmains, Volcanoguy, The Anomebot2, Deor, Mild Bill Hiccup, Resident Mario and Anonymous: 1

- **Eastern Gemini Seamount** *Source:* https://en.wikipedia.org/wiki/Eastern_Gemini_Seamount?oldid=683680608 *Contributors:* Gilgamesh~enwiki, D6, Grutness, Avenue, Jaraalbe, Droll, Volcanoguy, GeoWriter, SD Martin61, Lightbot, Resident Mario, Look2See1, Stormchaser89 and Op47

- **Explorer Seamount** *Source:* https://en.wikipedia.org/wiki/Explorer_Seamount?oldid=350013923 *Contributors:* D6, SmackBot, Skookum1, Volcanoguy, Sfan00 IMG and Plastikspork

- **Ferrel Seamount** *Source:* https://en.wikipedia.org/wiki/Ferrel_Seamount?oldid=536492874 *Contributors:* Bender235, Grutness, Woohookitty, Lightbot, Resident Mario and Jesse V.

- **Filippo Reef** *Source:* https://en.wikipedia.org/wiki/Filippo_Reef?oldid=663357040 *Contributors:* Enochlau, Beland, Thorwald, D6, Grutness, Avenue, Deror avi, Ratzer, Jaraalbe, Nicke L, SmackBot, MalafayaBot, Bardsandwarriors, Fenix down, Adolphus79, Goustien, MacedonianBoy, Addbot, TechBot, PigFlu Oink, EmausBot, ZéroBot, WPSamson, Helpful Pixie Bot, MozzazzoM, BGCP1 and Anonymous: 8

- **Foundation Seamounts** *Source:* https://en.wikipedia.org/wiki/Foundation_Seamounts?oldid=662505729 *Contributors:* Stemonitis, Daderot, Nerfer, Volcanoguy, Plastikspork and Addbot

- **Graham Seamount** *Source:* https://en.wikipedia.org/wiki/Graham_Seamount?oldid=617405260 *Contributors:* RedWolf, D6, Grutness, SmackBot, Volcanoguy, Inks.LWC, Sfan00 IMG, Plastikspork, Crusoe8181, Zingkeel and Anonymous: 1

- **Green Seamount** *Source:* https://en.wikipedia.org/wiki/Green_Seamount?oldid=678836553 *Contributors:* Mandarax, Rjwilmsi, CmdrObot, Rcej, Piledhigheranddeeper, Lightbot, Yobot, Citation bot, Resident Mario, Trappist the monk, Helpful Pixie Bot, Bibcode Bot, George Ponderevo, Anrnusna and Monkbot

- **Guide Seamount** *Source:* https://en.wikipedia.org/wiki/Guide_Seamount?oldid=584063653 *Contributors:* Rich Farmbrough, Woohookitty, Malcolma, Ken Gallager, Signalhead, WOSlinker, Iohannes Animosus, Lightbot, LilHelpa and Resident Mario

- **Gumdrop Seamount** *Source:* https://en.wikipedia.org/wiki/Gumdrop_Seamount?oldid=575501506 *Contributors:* Woohookitty, Ser Amantio di Nicolao, Igodard, Signalhead, Lightbot, Resident Mario, Crusoe8181, Escapepea, Khazar2 and Anonymous: 1

- **Heck Seamount** *Source:* https://en.wikipedia.org/wiki/Heck_Seamount?oldid=306739858 *Contributors:* D6, Grutness, SmackBot, Volcanoguy, Sfan00 IMG and Plastikspork

- **Schmieder Bank** *Source:* https://en.wikipedia.org/wiki/Schmieder_Bank?oldid=532298646 *Contributors:* Gary, SmackBot, Hmains, Colonies Chris, MarshBot, The Anomebot2, Schmieder, J.delanoy, Keesiewonder, Speciate, Lightmouse, RexxS, Lightbot, Gerixau, AnomieBOT, Look2See1, BG19bot and Anonymous: 4

- **Seminole Seamount** *Source:* https://en.wikipedia.org/wiki/Seminole_Seamount?oldid=306739842 *Contributors:* D6, Grutness, SmackBot, Volcanoguy, Sfan00 IMG, Plastikspork and Otolemur crassicaudatus

- **Siletz River Volcanics** *Source:* https://en.wikipedia.org/wiki/Siletz_River_Volcanics?oldid=609971001 *Contributors:* Gilgamesh~enwiki, Vsmith, Rjwilmsi, EncMstr, Droll, Cydebot, Mattisse, Volcanoguy, The Anomebot2, Kktor, J. Johnson, Chris.urs-o, RjwilmsiBot and Look2See1

- **South Chamorro Seamount** *Source:* https://en.wikipedia.org/wiki/South_Chamorro_Seamount?oldid=646942901 *Contributors:* Mandarax, BD2412, Jimp, Peter Horn, EoGuy, SchreiberBike, Yobot, EchetusXe, Resident Mario, Abductive, Jonesey95 and Anonymous: 1

- **Stirni Seamount** *Source:* https://en.wikipedia.org/wiki/Stirni_Seamount?oldid=306739851 *Contributors:* D6, Grutness, SmackBot, Volcanoguy, Sfan00 IMG, Plastikspork and Otolemur crassicaudatus

- **Suiyo Seamount** *Source:* https://en.wikipedia.org/wiki/Suiyo_Seamount?oldid=496734209 *Contributors:* Wetman, Klemen Kocjancic, Vsmith, GeeJo, Stepheng3, Lightbot, Citation bot, Resident Mario, FrescoBot, John of Reading and RockMagnetist

- **Supply Reef** *Source:* https://en.wikipedia.org/wiki/Supply_Reef?oldid=639322828 *Contributors:* Docu, RedWolf, D6, Darwinek, Grutness, Jaraalbe, Hmains, Droll, GVP Webmaster, Myasuda, Riffle, VolkovBot, Seattle Skier, Dominictimms, Goustien, Plastikspork, Detroiterbot, Addbot, Lightbot, Stormchaser89, ZéroBot, SporkBot and Anonymous: 3

- **Taney Seamounts** *Source:* https://en.wikipedia.org/wiki/Taney_Seamounts?oldid=588548777 *Contributors:* Wetman, DragonflySixtyseven, Shogun~enwiki, Pauli133, Woohookitty, Rwalker, Georgewilliamherbert, Moonriddengirl, Awickert, Iohannes Animosus, Lightbot, Resident Mario, Calmer Waters, Helpful Pixie Bot, BattyBot and Anonymous: 2

- **Tasmanian Seamounts** *Source:* https://en.wikipedia.org/wiki/Tasmanian_Seamounts?oldid=592043859 *Contributors:* Llywrch, Woohookitty, JEH, WolfmanSF, The Anomebot2, Sun Creator, Pgallert, Addbot, Neilgreatorex, Resident Mario, Jonesey95, EmausBot, BattyBot and Anonymous: 1

- **Teahitia** *Source:* https://en.wikipedia.org/wiki/Teahitia?oldid=543421757 *Contributors:* Gilgamesh~enwiki, Ratzer, FlaBot, Volcanoguy, The Anomebot2, TXiKiBoT, Plastikspork and Addbot

- **Three Wise Men (volcanoes)** *Source:* https://en.wikipedia.org/wiki/Three_Wise_Men_(volcanoes)?oldid=615590413 *Contributors:* Woohookitty, Rjwilmsi, Goustien, Mild Bill Hiccup, Resident Mario, Sopher99, Gaarmyvet and Anonymous: 1

- **Tucker Seamount** *Source:* https://en.wikipedia.org/wiki/Tucker_Seamount?oldid=306739833 *Contributors:* D6, Grutness, SmackBot, Volcanoguy, Sfan00 IMG, Plastikspork and Otolemur crassicaudatus

- **Tuzo Wilson Seamounts** *Source:* https://en.wikipedia.org/wiki/Tuzo_Wilson_Seamounts?oldid=617197177 *Contributors:* Dino, D6, Jaraalbe, Myasuda, Volcanoguy, Plastikspork, Full-date unlinking bot and Jo-Jo Eumerus

- **Union Seamount** *Source:* https://en.wikipedia.org/wiki/Union_Seamount?oldid=306739848 *Contributors:* D6, Grutness, SmackBot, Volcanoguy, Sfan00 IMG, Plastikspork and Otolemur crassicaudatus

- **Vailulu'u** *Source:* https://en.wikipedia.org/wiki/Vailulu'u?oldid=679110511 *Contributors:* Saforrest, Wwoods, Gilgamesh~enwiki, D6, Vsmith, Avenue, Johntex, Kitch, Tabletop, Rjwilmsi, Vegaswikian, Wahoo6942, Jaraalbe, Peter Grey, Poulpy, SmackBot, Droll, G716, GVP Webmaster, Btgiles, Cydebot, Volcanoguy, Nyttend, Daarznieks, DerHexer, Trusilver, GeoWriter, Hugo999, Goustien, Addbot, Lightbot, Peko, TheoryDerailed, Teinesavaii, Lotje, EmausBot, John of Reading, Δ, ClueBot NG, CactusBot, BattyBot, Makecat-bot, Monkbot and Anonymous: 8

- **Vance Seamounts** *Source:* https://en.wikipedia.org/wiki/Vance_Seamounts?oldid=614642879 *Contributors:* Grutness, Gene Nygaard, Woohookitty, Jaraalbe, Rwalker, SmackBot, Chris the speller, Courcelles, Volcanoguy, Addbot, Lightbot, Resident Mario and BattyBot

- **Winslow Reef, Phoenix Islands** *Source:* https://en.wikipedia.org/wiki/Winslow_Reef%2C_Phoenix_Islands?oldid=605609545 *Contributors:* Wwoods, Jason Quinn, D6, Eyrian, Grutness, Avenue, Deror avi, Ratzer, Jaraalbe, Renesis, Hmains, Bluebot, Pustelnik, Satori Son, Widders, Goustien, DragonBot, MacedonianBoy, Kcstover, Addbot, RevelationDirect, EmausBot, ChuispastonBot, Unscintillating, MozzazzoM, BattyBot and Anonymous: 5

- **Adare Seamounts** *Source:* https://en.wikipedia.org/wiki/Adare_Seamounts?oldid=601795560 *Contributors:* D6, Grutness, Caerwine, Hmains, Skizzik, AlbertHerring, JaGa, Hugo999, John Carter, Jan1nad, Yobot and Resident Mario

- **Balleny Seamounts** *Source:* https://en.wikipedia.org/wiki/Balleny_Seamounts?oldid=473709119 *Contributors:* Grutness, Hmains, Ser Amantio di Nicolao, Dr. Blofeld, Jan1nad and Kumi-Taskbot

- **Barsukov Seamount** *Source:* https://en.wikipedia.org/wiki/Barsukov_Seamount?oldid=557504453 *Contributors:* Grutness, Hmains, Dr. Blofeld, Jan1nad, Addbot and Kumi-Taskbot

- **Belgica Guyot** *Source:* https://en.wikipedia.org/wiki/Belgica_Guyot?oldid=519710905 *Contributors:* Grutness, Zntrip, Hmains, Dr. Blofeld, Jan1nad, Kumi-Taskbot and Elliechops

- **Boomerang Seamount** *Source:* https://en.wikipedia.org/wiki/Boomerang_Seamount?oldid=656987345 *Contributors:* D6, Grutness, Neilbeach, Hmains, Droll, Volcanoguy, Plasticup, Hugo999, Addbot, Lightbot, Peko, DARTH SIDIOUS 2, EmausBot, Ktmj97, Stormchaser89, ZéroBot, 4dedsel and Anonymous: 1

- **Christmas Island Seamount Province** *Source:* https://en.wikipedia.org/wiki/Christmas_Island_Seamount_Province?oldid=616335601 *Contributors:* The Anome, Wetman, Pigsonthewing, Vsmith, Mikenorton, Chienlit, WereSpielChequers, Goustien, Atethnekos, Resident Mario, RockMagnetist, CaroleHenson, Bibcode Bot, George Ponderevo, Avi8tor and Monkbot

- **Dallmann Seamount** *Source:* https://en.wikipedia.org/wiki/Dallmann_Seamount?oldid=475507142 *Contributors:* Grutness, Hmains, Ser Amantio di Nicolao, Hebrides, Dr. Blofeld, Jan1nad, Usb10 and Kumi-Taskbot

- **De Gerlache Seamounts** *Source:* https://en.wikipedia.org/wiki/De_Gerlache_Seamounts?oldid=551940203 *Contributors:* Grutness, Hmains, Dr. Blofeld, Jan1nad, Addbot and Kumi-Taskbot

- **Empedocles (volcano)** *Source:* https://en.wikipedia.org/wiki/Empedocles_(volcano)?oldid=687912532 *Contributors:* Jason Quinn, MacGyver-Magic, MakeRocketGoNow, Giraffedata, Grutness, Woohookitty, MZMcBride, NawlinWiki, Attilios, SmackBot, Hmains, TheKMan, Steff, Phaid, Sir marek, Frokor, Alaibot, Escarbot, Ilion2, Widefox, Otrebla86~enwiki, The Anomebot2, Plasticup, Stambouliote, STBotD, Seattle Skier, AlleborgoBot, Goustien, Plastikspork, Addbot, LilHelpa, Resident Mario, Chaiten1, RedBot, Eyadhamid and Anonymous: 8

- **Eratosthenes Seamount** *Source:* https://en.wikipedia.org/wiki/Eratosthenes_Seamount?oldid=541668079 *Contributors:* SetarconeX, Balcer, Grutness, Avenue, Jaraalbe, Hmains, Mikenorton, Volcanoguy, Hugo999, Gaianauta, F.chiodo, Addbot, Lightbot, Resident Mario, Mjbmrbot and ElphiBot

- **Graham Island (Sicily)** *Source:* https://en.wikipedia.org/wiki/Graham_Island_(Sicily)?oldid=687274890 *Contributors:* Joakim Ziegler, Danny, Michael Hardy, Etherialemperor, דוד, Joy, Ccady, Wetman, Chrism, Auric, Saforrest, Carnildo, Brecchie, Gilgamesh~enwiki, Jason Quinn, Eregli bob, Simhedges, Talrias, Blue387, Neutrality, Clawed, Cyclopia, Nabla, Worldtraveller, Shanes, Terrycojones, Anthony Appleyard, Axl, Pippu d'Angelo, Avenue, Shogun~enwiki, Gpvos, Stefanomione, Graham87, Rjwilmsi, Koavf, MZMcBride, Str1977, Seegoon, Pyrotec, Zwobot, Galar71, Scoutersig, Attilios, SmackBot, FocalPoint, Gnangarra, Thumperward, MaartenVidal, TheKMan, Phaid, Lambiam, Tk-tktk, Frokor, Floridan, Haus, Van helsing, Runningonbrains, Cydebot, Marcuscalabresus, Malleus Fatuorum, Thijs!bot, Widefox, Kayamon, Volcanoguy, Acroterion, Celithemis, The Anomebot2, Archolman, Nono64, Stambouliote, KylieTastic, STBotD, 1812ahill, Hugo999, TXiK-iBoT, Nedrutland, Tikuko, Ceranthor, AlleborgoBot, SieBot, Goustien, Lightmouse, Seuraza, Plastikspork, SkeletorUK, Babbaluci, Addbot, Debresser, SamatBot, Mps, Yobot, AnomieBOT, Citation bot, Xqbot, J04n, Resident Mario, Chris.urs-o, FrescoBot, Dawnnight, DrilBot, Chaiten1, Abductive, Pbsouthwood, Underlying lk, The grey side, TrueGrave13, ZéroBot, Eyadhamid, Chris857, Noble fan, SkepticalRaptor, YFdyh-bot, Lubiesque, Dexbot, Epicgenius, Gasmonitor, SkateTier, Tihkon2, Lesbianism is evil and Anonymous: 49

- **Hakurei Seamount** *Source:* https://en.wikipedia.org/wiki/Hakurei_Seamount?oldid=495117007 *Contributors:* Grutness, Hmains, Ser Amantio di Nicolao, Dr. Blofeld, Jan1nad and TPBot

- **Iselin Seamount** *Source:* https://en.wikipedia.org/wiki/Iselin_Seamount?oldid=501624555 *Contributors:* D6, Grutness, Hmains, Skizzik, Ser Amantio di Nicolao, Fleebo, Jan1nad and Addbot

- **Lecointe Guyot** *Source:* https://en.wikipedia.org/wiki/Lecointe_Guyot?oldid=558766491 *Contributors:* Grutness, Hmains, Dr. Blofeld, Cyfal, Jan1nad, Kumi-Taskbot, TPBot and Lemnaminor

- **Lichtner Seamount** *Source:* https://en.wikipedia.org/wiki/Lichtner_Seamount?oldid=559851098 *Contributors:* Grutness, Kam Solusar, Hmains, Dr. Blofeld, Jan1nad, Addbot and TPBot

- **Marsili** *Source:* https://en.wikipedia.org/wiki/Marsili?oldid=622441181 *Contributors:* Xezbeth, Thoken, Rjwilmsi, Mix321, Dan Pelleg, Paris1127, Rep07, Cryonic07, Addbot, Favonian, Lightbot, Luckas-bot, ArthurBot, Xqbot, Fabyno, H3llBot, 4dedsel, BG19bot and Anonymous: 4

- **Maud Seamount** *Source:* https://en.wikipedia.org/wiki/Maud_Seamount?oldid=570729339 *Contributors:* Grutness, Hmains, Ser Amantio di Nicolao, Katharineamy, Jan1nad, WikiCopter and TPBot

- **Muirfield Seamount** *Source:* https://en.wikipedia.org/wiki/Muirfield_Seamount?oldid=620510829 *Contributors:* Saforrest, Wwoods, D6, Avenue, Saberwyn, SmackBot, Peter Horn, Shikaga, BoH, Volcanoguy, BloodDoc, Hugo999, F.chiodo, Seuraza, Addbot, Lightbot, Agrophobe, DexDor, John of Reading, Canstusdis and Anonymous: 5

- **Orca Seamount** *Source:* https://en.wikipedia.org/wiki/Orca_Seamount?oldid=614802321 *Contributors:* Wwoods, YUL89YYZ, Grutness, Woohookitty, Mandarax, Ground Zero, Hmains, Lambiam, Apcbg, Gierszep, Stepheng3, Resident Mario and Locobot

- **Rosenthal Seamount** *Source:* https://en.wikipedia.org/wiki/Rosenthal_Seamount?oldid=494389168 *Contributors:* Grutness, Hmains, Dr. Blofeld and TPBot

- **Walters Shoals** *Source:* https://en.wikipedia.org/wiki/Walters_Shoals?oldid=683745081 *Contributors:* Palnatoke, Hike395, Stepp-Wulf, Grutness, Hmains, Marek69, Hugo999, Gbawden, Mild Bill Hiccup, Stepheng3, Lightbot, ONaNcle, The High Fin Sperm Whale, Lotje and Anonymous: 2

- **Wordie Seamount** *Source:* https://en.wikipedia.org/wiki/Wordie_Seamount?oldid=550653707 *Contributors:* Apcbg

- **Rio Grande Rise** *Source:* https://en.wikipedia.org/wiki/Rio_Grande_Rise?oldid=670994051 *Contributors:* Topbanana, Waltpohl, Dawnseeker2000, Fama Clamosa and Glevum

- **Walvis Ridge** *Source:* https://en.wikipedia.org/wiki/Walvis_Ridge?oldid=687703389 *Contributors:* Rjwilmsi, Bgwhite, Willy turner, Clarityfiend, Rosarinagazo, Dawnseeker2000, Volcanoguy, Tonicthebrown, R'n'B, Addbot, DougsTech, Luckas-bot, Elm-39, Gigemag76, GrouchoBot, Chris.urs-o, Fama Clamosa, RjwilmsiBot, Bibcode Bot, Glevum, Jami430, RichardMills65, Lappspira and Anonymous: 3

7.2   Images

- **File:04Sep2007_Etna_from_SE_Crater.jpg** *Source:* https://upload.wikimedia.org/wikipedia/commons/8/80/04Sep2007_Etna_from_SE_Crater.jpg *License:* CC BY 3.0 *Contributors:* Own work *Original artist:* Jason Bott, Christopher Berger, Pete Garza

- **File:Adelie_stub_map.png** *Source:* https://upload.wikimedia.org/wikipedia/en/3/39/Adelie_stub_map.png *License:* CC-BY-SA-2.5 *Contributors:* ? *Original artist:* ?

- **File:Ambox_important.svg** *Source:* https://upload.wikimedia.org/wikipedia/commons/b/b4/Ambox_important.svg *License:* Public domain *Contributors:* Own work, based off of Image:Ambox scales.svg *Original artist:* Dsmurat (talk · contribs)

- **File:Ile_Barren,_1995.jpg** *Source:* https://upload.wikimedia.org/wikipedia/commons/b/b3/Ile_Barren%2C_1995.jpg *License:* Public domain *Contributors:* https://archive.org/details/STS067-721A-052 *Original artist:* NASA

- **File:Italy_provincial_location_map.svg** *Source:* https://upload.wikimedia.org/wikipedia/commons/6/6b/Italy_provincial_location_map.svg *License:* CC BY-SA 3.0 *Contributors:* This vector graphics image was created with Adobe Illustrator. *Original artist:* TUBS* <a href='//commons. wikimedia.org/wiki/User_talk:TUBS' title='User talk:TUBS'><img alt='Email Silk.svg' src='https://upload.wikimedia.org/wikipedia/commons/ thumb/5/5f/Email_Silk.svg/15px-Email_Silk.svg.png' width='15' height='15' srcset='https://upload.wikimedia.org/wikipedia/commons/thumb/ 5/5f/Email_Silk.svg/23px-Email_Silk.svg.png 1.5x, https://upload.wikimedia.org/wikipedia/commons/thumb/5/5f/Email_Silk.svg/30px-Email_ Silk.svg.png 2x' data-file-width='16' data-file-height='16' /></a> <a href='http://toolserver.org/~{}daniel/WikiSense/Gallery.php?wikifam= commons.wikimedia.org,,&,,img_user_text=TUBS' title='Gallery' data-x-rel='nofollow'><img alt='Gallery' src='https://upload. wikimedia.org/wikipedia/commons/thumb/5/59/Gnome-emblem-photos.svg/20px-Gnome-emblem-photos.svg.png' width='20' height='20' src- set='https://upload.wikimedia.org/wikipedia/commons/thumb/5/59/Gnome-emblem-photos.svg/30px-Gnome-emblem-photos.svg.png 1.5x, https://upload.wikimedia.org/wikipedia/commons/thumb/5/59/Gnome-emblem-photos.svg/40px-Gnome-emblem-photos.svg.png 2x' data-file- width='48' data-file-height='48' /></a>

- **File:KI_Line_islands.PNG** *Source:* https://upload.wikimedia.org/wikipedia/commons/4/43/KI_Line_islands.PNG *License:* Public domain *Contributors:* Own work *Original artist:* Hobe / Holger Behr

- **File:KermadecIsland.A2012225.2140.250m.jpg** *Source:* https://upload.wikimedia.org/wikipedia/commons/5/5b/KermadecIsland.A2012225. 2140.250m.jpg *License:* Public domain *Contributors:* http://rapidfire.sci.gsfc.nasa.gov/cgi-bin/imagery/single.cgi?image=KermadecIsland. A2012225.2140.250m.jpg *Original artist:* NASA

- **File:Kermadec_Arc.jpg** *Source:* https://upload.wikimedia.org/wikipedia/commons/d/d4/Kermadec_Arc.jpg *License:* Public domain *Con- tributors:* http://oceanexplorer.noaa.gov/explorations/05fire/background/plan/media/sw_pac_legs.html *Original artist:* ?

- **File:Kodiak-Bowie_Seamounts.jpg** *Source:* https://upload.wikimedia.org/wikipedia/commons/5/50/Kodiak-Bowie_Seamounts.jpg *License:* Public domain *Contributors:* http://www.oceanexplorer.noaa.gov/explorations/04alaska/background/volcanic/media/gofae03_map.html *Orig- inal artist:* ?

- **File:Koral1.jpg** *Source:* https://upload.wikimedia.org/wikipedia/commons/f/fc/Koral1.jpg *License:* CC-BY-SA-3.0 *Contributors:* No machine- readable source provided. Own work assumed (based on copyright claims). *Original artist:* No machine-readable author provided. Kluka assumed (based on copyright claims).

- **File:Koryaksky_volcano_Petropavlovsk-Kamchatsky_oct-2005.jpg** *Source:* https://upload.wikimedia.org/wikipedia/commons/e/e0/Koryaksky_ volcano_Petropavlovsk-Kamchatsky_oct-2005.jpg *License:* Public domain *Contributors:* ? *Original artist:* ?

- **File:Lakagigar_Iceland_2004-07-01.jpg** *Source:* https://upload.wikimedia.org/wikipedia/commons/1/13/Lakagigar_Iceland_2004-07-01. jpg *License:* CC-BY-SA-3.0 *Contributors:* Own work (self-made photo) *Original artist:* Juhász Péter

- **File:Lava_Lake_Nyiragongo_2.jpg** *Source:* https://upload.wikimedia.org/wikipedia/commons/7/76/Lava_Lake_Nyiragongo_2.jpg *License:* CC BY-SA 3.0 *Contributors:* Own work *Original artist:* User:Cai Tjeenk Willink (Caitjeenk)

- **File:Lava_channel_overflow.JPG** *Source:* https://upload.wikimedia.org/wikipedia/commons/4/41/Lava_channel_overflow.JPG *License:* CC BY-SA 3.0 *Contributors:* Own work *Original artist:* Brocken Inaglory

- **File:Lava_entering_sea_-_Hawaii.png** *Source:* https://upload.wikimedia.org/wikipedia/commons/8/8c/Lava_entering_sea_-_Hawaii.png *License:* CC BY-SA 2.0 *Contributors:* Lava 7 *Original artist:* Jennifer Williams from Hayward, USA

- **File:LocationAmericanSamoa.png** *Source:* https://upload.wikimedia.org/wikipedia/commons/3/33/LocationAmericanSamoa.png *License:* Public domain *Contributors:* ? *Original artist:* Location map for the American Samoa.

- **File:LocationPhoenix.png** *Source:* https://upload.wikimedia.org/wikipedia/commons/1/1a/LocationPhoenix.png *License:* CC BY-SA 2.5 *Contributors:* No machine-readable source provided. Own work assumed (based on copyright claims). *Original artist:* No machine-readable author provided. Kmusser assumed (based on copyright claims).

- **File:Location_of_Ferdinandea.jpg** *Source:* https://upload.wikimedia.org/wikipedia/commons/7/70/Location_of_Ferdinandea.jpg *License:* CC-BY-SA-3.0 *Contributors:* ? *Original artist:* ?

- **File:LoihiBathemetric.jpg** *Source:* https://upload.wikimedia.org/wikipedia/commons/9/99/LoihiBathemetric.jpg *License:* Public domain *Contributors:* Origin *Original artist:* NOAA

- **File:Loihi_3d.gif** *Source:* https://upload.wikimedia.org/wikipedia/commons/4/49/Loihi_3d.gif *License:* Public domain *Contributors:* (Origi- nal text: National Science Foundation / National Oceanic and Atmospheric Administration / The Hawaii Undersea Research Laboratory (HURL) *http://www.soest.hawaii.edu/HURL/images/loihi_3d.gif http://www.soest.hawaii.edu/HURL/hurl_loihi.html*) Original artist: *John Smith and Brooks Bays*

- **File:Loihiflank.jpg** *Source:* https://upload.wikimedia.org/wikipedia/commons/0/00/Loihiflank.jpg *License:* Public domain *Contributors:* http: //www.photolib.noaa.gov/htmls/nur05020.htm. Transferred from en.wikipedia to Commons by PatríciaR. *Original artist:* ?

- **File:Louisville_seamount_chain_-_bathymetry.jpg** *Source:* https://upload.wikimedia.org/wikipedia/commons/c/c1/Louisville_seamount_ chain_-_bathymetry.jpg *License:* Public domain *Contributors:* Ultimate source: ETOPO2v2, a digital database of seafloor and land elevations. Actually cropped from Image:Pacific_elevation.jpg. *Original artist:* World Data Center for Geophysics & Marine Geology (Boulder, CO), National Geophysical Data Center, NOAA

- **File:Maldivesfish2.jpg** *Source:* https://upload.wikimedia.org/wikipedia/commons/3/35/Maldivesfish2.jpg *License:* CC BY-SA 2.0 *Contrib- utors:* Originally uploaded to **Flickr** as Fishes *Original artist:* Betty x1138

- **File:Map_of_Graham_Island.jpg** *Source:* https://upload.wikimedia.org/wikipedia/commons/d/dc/Map_of_Graham_Island.jpg *License:* CC BY-SA 3.0 *Contributors:* Paint *Original artist:* Gasmonitor

- **File:Mauna_Kea_from_the_ocean.jpg** *Source:* https://upload.wikimedia.org/wikipedia/commons/8/8d/Mauna_Kea_from_the_ocean.jpg *License:* CC BY 2.0 *Contributors:* originally posted to **Flickr** as IMG_2673.JPG *Original artist:* Vadim Kurland

- **File:Mauna_Loa_atmospheric_transmission.png** *Source:* https://upload.wikimedia.org/wikipedia/commons/9/9c/Mauna_Loa_atmospheric_transmission.png *License:* Public domain *Contributors:* ? *Original artist:* ?

- **File:MtCleveland_ISS013-E-24184.jpg** *Source:* https://upload.wikimedia.org/wikipedia/commons/4/4a/MtCleveland_ISS013-E-24184.jpg *License:* Public domain *Contributors:* http://earthobservatory.nasa.gov/Newsroom/NewImages/images.php3?img_id=17285 *Original artist:* ISS Crew Earth Observations experiment and the Image Science & Analysis Group, Johnson Space Center.

- **File:NE_seamounts.jpg** *Source:* https://upload.wikimedia.org/wikipedia/commons/f/f7/NE_seamounts.jpg *License:* Public domain *Contributors:* http://www.oceanexplorer.noaa.gov/explorations/03mountains/background/plan/media/sites.html *Original artist:* NOAA

- **File:N_Atlantic_seamounts_(Converted).pdf** *Source:* https://upload.wikimedia.org/wikipedia/commons/5/5a/N_Atlantic_seamounts_%28Converted%29.pdf *License:* CC BY-SA 4.0 *Contributors:* Own work *Original artist:* PeterTHarris

- **File:Narcondam_island.jpg** *Source:* https://upload.wikimedia.org/wikipedia/commons/2/27/Narcondam_island.jpg *License:* GFDL 1.2 *Contributors:* ? *Original artist:* ?

- **File:NewEngland_Seamount_Chain.jpg** *Source:* https://upload.wikimedia.org/wikipedia/commons/5/58/NewEngland_Seamount_Chain.jpg *License:* Public domain *Contributors:* ? *Original artist:* ?

- **File:New_England_Seamount_community.jpg** *Source:* https://upload.wikimedia.org/wikipedia/commons/c/c6/New_England_Seamount_community.jpg *License:* Public domain *Contributors:* Source: [1]. *Original artist:* ?

- **File:Newfoundland_Seamounts.jpg** *Source:* https://upload.wikimedia.org/wikipedia/commons/8/84/Newfoundland_Seamounts.jpg *License:* Public domain *Contributors:* NASA World Wind *Original artist:* NASA

- **File:Ocean_Bottom_Observatory_at_Pele'{}s_Vent.jpg** *Source:* https://upload.wikimedia.org/wikipedia/commons/2/21/Ocean_Bottom_Observatory_at_Pele%27s_Vent.jpg *License:* Public domain *Contributors:* [1]Origin *Original artist:* NOAA

- **File:Oceania_(orthographic_projection).svg** *Source:* https://upload.wikimedia.org/wikipedia/commons/8/8e/Oceania_%28orthographic_projection%29.svg *License:* CC BY-SA 3.0 *Contributors:* Based on File:Australia (orthographic projection).svg *Original artist:* Ch1902

- **File:Olympus_Mons.jpeg** *Source:* https://upload.wikimedia.org/wikipedia/commons/7/75/Olympus_Mons.jpeg *License:* Public domain *Contributors:* ? *Original artist:* ?

- **File:Orange_roughy.png** *Source:* https://upload.wikimedia.org/wikipedia/commons/2/2e/Orange_roughy.png *License:* Public domain *Contributors:* ? *Original artist:* ?

- **File:Orca_LocationMap.jpg** *Source:* https://upload.wikimedia.org/wikipedia/en/2/20/Orca_LocationMap.jpg *License:* CC-BY-2.5 *Contributors:*

[1] *Original artist:*

www.geody.com, satellite mapping

- **File:OregonCoastRangeRocks.JPG** *Source:* https://upload.wikimedia.org/wikipedia/commons/4/4f/OregonCoastRangeRocks.JPG *License:* Public domain *Contributors:* Transferred from en.wikipedia to Commons by Aboutmovies using CommonsHelper. *Original artist:* M.O. Stevens at en.wikipedia

- **File:Orthographic_projection_centred_over_Easter_Island.png** *Source:* https://upload.wikimedia.org/wikipedia/commons/a/a7/Orthographic_projection_centred_over_Easter_Island.png *License:* Public domain *Contributors:* ? *Original artist:* ?

- **File:PD-icon.svg** *Source:* https://upload.wikimedia.org/wikipedia/en/6/62/PD-icon.svg *License:* PD *Contributors:* ? *Original artist:* ?

- **File:People_icon.svg** *Source:* https://upload.wikimedia.org/wikipedia/commons/3/37/People_icon.svg *License:* CC0 *Contributors:* OpenClipart *Original artist:* OpenClipart

- **File:Peter_Mel_at_Cortez_Bank.jpg** *Source:* https://upload.wikimedia.org/wikipedia/commons/b/b8/Peter_Mel_at_Cortez_Bank.jpg *License:* Public domain *Contributors:* Own work *Original artist:* PPNF

- **File:Pierce_Seamount.jpg** *Source:* https://upload.wikimedia.org/wikipedia/commons/0/0a/Pierce_Seamount.jpg *License:* Public domain *Contributors:* NASA World Wind *Original artist:* NASA

- **File:Pillow_basalt_crop_l.jpg** *Source:* https://upload.wikimedia.org/wikipedia/commons/0/08/Pillow_basalt_crop_l.jpg *License:* Public domain *Contributors:* Corrected version of Nur05018.jpg (taken from http://www.photolib.noaa.gov/htmls/nur05018.htm) *Original artist:* National Oceanic and Atmospheric Administration

- **File:Pinatubo_ash_plume_910612.jpg** *Source:* https://upload.wikimedia.org/wikipedia/commons/1/1f/Pinatubo_ash_plume_910612.jpg *License:* Public domain *Contributors:* http://vulcan.wr.usgs.gov/Volcanoes/Philippines/Pinatubo/images.html *Original artist:* D. Harlow

- **File:Pisces_V.jpg** *Source:* https://upload.wikimedia.org/wikipedia/commons/4/48/Pisces_V.jpg *License:* Public domain *Contributors:* [1]Origin *Original artist:* NOAA

- **File:Pompeii_-_Casa_del_Centenario_-_MAN.jpg** *Source:* https://upload.wikimedia.org/wikipedia/commons/b/bf/Pompeii_-_Casa_del_Centenario_-_MAN.jpg *License:* Public domain *Contributors:* Marisa Ranieri Panetta (ed.): Pompeji. Geschichte, Kunst und Leben in der versunkenen Stadt. Belser, Stuttgart 2005, ISBN 3-7630-2266-X, p. 111 *Original artist:* WolfgangRieger

- **File:Portal-puzzle.svg** *Source:* https://upload.wikimedia.org/wikipedia/en/f/fd/Portal-puzzle.svg *License:* Public domain *Contributors:* ? *Original artist:* ?

- **File:Puu_Oo_cropped.jpg** *Source:* https://upload.wikimedia.org/wikipedia/commons/6/6d/Puu_Oo_cropped.jpg *License:* Public domain *Contributors:* USGS *Original artist:* G.E. Ulrich, USGS. Cropping by Hike395 (talk · contribs)

- **File:Waves_in_pacifica_1.jpg** *Source:* https://upload.wikimedia.org/wikipedia/commons/4/45/Waves_in_pacifica_1.jpg *License:* GFDL *Contributors:* Own work *Original artist:* Brocken Inaglory

- **File:WikiProject_Geology.svg** *Source:* https://upload.wikimedia.org/wikipedia/commons/e/e7/WikiProject_Geology.svg *License:* CC BY-SA 2.5 *Contributors:* ? *Original artist:* ?

- **File:Wiki_letter_w.svg** *Source:* https://upload.wikimedia.org/wikipedia/en/6/6c/Wiki_letter_w.svg *License:* Cc-by-sa-3.0 *Contributors:* ? *Original artist:* ?

- **File:Wikivoyage-Logo-v3-icon.svg** *Source:* https://upload.wikimedia.org/wikipedia/commons/d/dd/Wikivoyage-Logo-v3-icon.svg *License:* CC BY-SA 3.0 *Contributors:* Own work *Original artist:* AleXXw

- **File:Zavodovski-Location.JPG** *Source:* https://upload.wikimedia.org/wikipedia/commons/7/79/Zavodovski-Location.JPG *License:* CC-BY-SA-3.0 *Contributors:* ? *Original artist:* ?

7.3   Content license